WITHDRAWN

INTRODUCTION
TO TIDES

Low Tide in Minas Basin. Photo by Miss Mildred E. Collins.

INTRODUCTION TO TIDES

The Tides of the Waters of
New England and New York

Alfred C. Redfield

MARINE SCIENCE INTERNATIONAL
Woods Hole, Massachusetts

Cover photo by
Elizabeth Fye
Woods Hole Oceanographic Institution

Copyright © 1980 by Alfred Clarence Redfield
Library of Congress Catalog Card Number: 81-13365
ISBN: 0-87933-901-2

All rights reserved. No part of this book may be reproduced or transmitted in any form or by any means—graphic, electronic, or mechanical, including photocopying, recording, taping, or information storage and retrieval systems—without written permission of the publisher.

83 82 81   1 2 3 4 5
Manufactured in the United States of America.

Library of Congress Cataloging in Publication Data

Redfield, Alfred Clarence, 1890-
    Introduction to tides.
    Originally published: The tides of the waters of
New England and New York. Woods Hole, Mass.: Woods Hole
Oceanographic Institution, 1980.
    Includes bibliographical references.
    1. Tides—New England. 2. Tides—New York (State)
I. Title.
GC358.N48R42 1981   551.47′08′0974   81-13365
ISBN 0-87933-901-2   AACR2

Dedicated to the members of the United States Coast and Geodetic Survey whose labor has made this book possible.

FOREWORD

Fifteen years ago Alfred Redfield's friends, associates and students prepared a volume of scientific papers in his honor on his 75th birthday. It is characteristic of the man that today, as he approaches his 90th birthday, this volume on **The Tides of the Waters of New England and New York** flows from his own pen. His interest in natural phenomena and his curiosity about them have been undiminished by passing years — indeed these years have given him the wisdom and insight to understand and to explain complex problems.

Alfred Redfield's interests have always been very broad, and he has made significant scientific contributions to many fields of knowledge. After earning a worldwide reputation in physiology, particularly with regard to muscle physiology, the effects of ionizing radiation on organisms and the gas exchanges of the blood, especially hemocyanin, he devoted more and more time to one of his first loves — life in the sea. He received one of the first appointments, as Senior Biologist, to the scientific staff of the fledgling Woods Hole Oceanographic Institution which is celebrating its 50th anniversary this year.

Alfred Redfield has a holistic approach to the biology of the sea, with the organisms drawing their sustenance from the nutrients dissolved in the water and returning these by excretion, death and decay in the endless cycle of life in the sea. One of his earliest marine publications, in 1934, was "On the proportions of organic derivatives in sea water and their relation to the composition of the plankton".[a] Redfield brought order out of chaos by reorganizing the consistency of the ratios of essential nutrient elements as they are assimilated by the phytoplankton and later returned to the water by regeneration. This consistency, not previously recognized, forms the cornerstone of modern studies of the dynamics of phytoplankton populations.

Alfred Redfield's interest in the tides dates, no doubt, to his boyhood when he dug for clams on the flats of Barnstable Harbor and sailed his boats beset by the tidal currents common in Cape Cod waters. In 1950, he published his first definitive paper on "The analysis of tidal phenomena in narrow embayments",[b] an analysis which he further develops and expands

[a] Redfield, A. C., 1934. pp. 176-192. In: James Johnstone, Memorial Volume, University of Liverpool Press.

[b] Redfield, A. C., 1950. Papers in Physical Oceanography and Meteorology 11(4): 36 pp.

in this volume. While others have written about the ocean tides, Redfield treats the inshore phenomena where tides and tidal currents are important to man's activities. From the predictions of the time and height of the tide and the velocity and direction of the related tidal currents, Redfield explains the effects of the reflected wave and the influence of adjacent topographic boundaries so that the *reasons* behind the observed phenomena can be understood.

One of Redfield's important characteristics is his ability to focus on critical problems or mechanisms. He has a unique ability to explain complex problems in terms understandable by both novice and specialist. He does not eschew mathematics but recognizes it as a useful tool for the manipulation of information in recognized and well-established ways. As he states in the introduction to this volume, "The important thing is to understand the *ideas* on which the mathematics is based, and these can be stated in ordinary language". All too often scientists obscure their thoughts in a complex language understood only by their close associates; but Alfred Redfield has always felt that it was not only essential to understand a problem, but also to explain it in a straightforward way so that it could be equally understood by the reader.

All who know Alfred Redfield admire him not only for his competence and his integrity, but also for his understanding and sympathy for those who follow in his footsteps. We welcome this volume which confirms his continuing productivity and interests in developing a more thorough understanding of the natural phenomena of the sea.

Bostwick H. Ketchum   April 1980

CONTENTS

Foreword		ix
Preface		xii
Chapter I.	The Origin of the Tide	1
Chapter II.	The Progressive Wave	6
Chapter III.	Nomographic Analysis—The Tide In Straits	13
Chapter IV.	The Reflected Co-oscillation of Embayments	19
Chapter V.	Hydraulic Currents	25
Chapter VI.	Shallow Water Tides and Harmonics	31
Chapter VII.	The Rotation of the Earth	38
Chapter VIII.	Meteorological Effects on the Tide	45
Chapter IX.	Sea Level	58
Chapter X.	The Tide Offshore	62
Chapter XI.	The Waters of Northern New England	67
Chapter XII.	The Waters of Southern New England	77
Chapter XIII.	New York Waters	89
Appendix A.	Observations and Experiments on Waves	103
Appendix B.	Conversion Factors for Units of Measurement and Useful Constants	108

PREFACE

This book is written for the many intelligent people who work or play along the coast between Sandy Hook and the Bay of Fundy in the hope that it will give them a better understanding of matters which greatly influence the daily ordering of their activities. It may be of value to the serious student of the tides, at the beginning as an introduction to tidal theory and later as a summary of the tides on this particular coast. The stretch of coast considered and the off-lying ocean contain examples of practically all known tidal phenomena.

The book is based for the most part on information given in the tide and current tables published by the U.S. Department of Commerce, National Oceanic and Atmospheric Administration, formerly the Coast and Geodetic Survey. It is not intended to replace these tables if one would know what to expect at any particular place on any particular day. Rather, it attempts to explain why the tide locally is as it is and why it varies from place to place.

Two sorts of information are given and are to be compared. The first are the predictions, based on observations, which are given in the tide and current tables. These are considered to be correct. The second are the results of theory.

The theory may be considered valid if its results correspond with the predictions. It must always be remembered that some other theory might give equally valid results. The most that can be hoped for is the demonstration that the theory on which the explanation of the tide is based is probable, not that is is a certainty. The theory of tides has been developed in mathematical terms and it is not possible to present it otherwise. Unfortunately, many of the intended readers have forgotten such knowledge of elementary mathematics as they may have had, or have had none at all. Such readers must take the mathematics for granted, realizing that mathematics is merely a way of putting ideas in the form of symbols which may then be manipulated in well-established ways. The important thing is to understand the *ideas* on which the mathematics is based and these can be stated in ordinary language.

In natural channels such as the bays and straits along the coast the depth and width are variable, and it cannot be assumed that progressive waves advance over equal distances in equal times. The important assumption on which the present analysis is based is that irregularities in the cross section of the channel do not change the relations at any position of elevation, time, and phase of the waves. This allows the tide at any place to be analyzed in terms of the phase of the progressive waves there present, but to be related to the geography only by the phase at that place. In other words, the character of the tide in any channel is determined from its observed behavior rather than any effect calculated from the dimensions of the channel.

The general theory by which the tides of this coast are explained is that the tide originates in the deep water of the off-lying ocean, principally as the result of the gravitational attraction of the moon. In crossing the continental shelf the progressive wave so produced is modified by interference of a wave arising from reflection from the coast. As the tide enters the straits and embayments which are tributary to the offshore waters, it is further modified; in the case of straits, by the interference of waves entering from their opposite ends; in that of embayments, by interference between the entering wave and its reflection from the head of the embayment. In both cases the tide is modified by the attenuation which these waves undergo within the passage and in the case of straits by the relative amplitude of the waves which enter from opposite ends. The theoretical treatment allows these effects to be evaluated numerically and thus indicates how the differences in behavior of the tide in these passages are to be explained.

The first two chapters consider the origin of the tide, the constituents or partial tides of which it is composed and the properties of the progressive wave, which is the form these constituents are assumed to have. The next two chapters deal with the interference which results when two progressive waves occur simultaneously in the same channel. Chapter III considers the case in straits which leads to the development of theoretical equations generally applicable to and of use in testing the theory. Chapter IV deals with the special case of embayments in which interference occurs between the entering wave and its reflection. The following five chapters consider special conditions which may cause the predicted tide to differ in certain respects from those predicted by the theoretical equations. These nine chapters are general in their application. They may be expected to hold irrespective of the location of the positions on the coast provided suitable conditions exist for their application. The final four chapters review the tides in geographically distinct regions. Some reader may turn first to the chapter on the tides of the region in which the place of his particular interest lies. If he does so, he may need to turn back to earlier chapters in order to understand what he is reading.

Appendix A contains an account of some observations and experiments which the reader may make for himself. If he will take the trouble to do so, matters described in Chapter II will become more real.

In order to have a common measure of time, the time of high-water, slack water, or other events is expressed relative to the time of the moon's transit at Greenwich, the Greenwich interval. Thus the *epoch* or *phase lag* of a constituent of the tide may be expressed relative to the time of transit at Greenwich of the body responsible for it, and it is known as the *Greenwich Epoch*.

Measurements of length are given in English units of feet and nautical miles, these being the units used in the tide and current tables and on nautical charts. Appendix B contains conversion factors from which metric equivalents may be obtained.

At the end of each chapter references are given which have been useful to the author and may interest those who wish to look into the matter further. Books that treat the tide in general in a nontechnical manner are:

Marmer, H.A. The Tide. Appleton, New York and London, 282 pp., 1926.
Defant, Albert. Ebb and Flow. University of Michigan, Ann Arbor, 121 pp., 1958
Darwin, G. H. The Tides. Freeman, San Francisco and New York, 387 pp., 1962.
Clancy, E. P. The Tides. Doubleday, Garden City, 228 pp., 1968.

Acknowledgements. The author wishes to express his indebtedness to those who have assisted in the preparation of this book, especially to Henry M. Stommel who developed the theoretical equations given in Chapter II which have been found to be applicable to many straits and embayments, to William S. von Arx who has verified parts of the text, to John A. Moody who supplied the records of the tide at North Falmouth, Falmouth Harbor, and Bournes Pond, to R. C. Beardsley, David Greenberg, W. R. Wright, and the National Oceanic and Atmospheric Administration for unpublished information. Mrs. Frances Dunlap has prepared most of the illustrations.

Chapter I

THE ORIGIN OF THE TIDE

The tide is due to small differences which result from the earth's rotation in the gravitation exerted by the heavenly bodies on the earth as a whole and on the water on its surface. It may be shown that only the moon and the sun produce sufficient force on the earth and its water to cause a measurable tide, the other planets being too small or too distant to cause a significant effect. As the earth rotates, however, the distance between the water at a point on its surface and the moon or sun varies slightly. The small variation in this distance produces the variation of gravitational force on the water on which their contribution to the tide depends.

The gravitational forces due to the moon and sun tend to move the water toward the position immediately underlying the body. They act tangentially to the earth's surface and are known as *tractive forces*. As the earth rotates, the strength and direction of the tractive forces change. When a particle of water is at a position where the moon rises or sets, there is no difference between the moon's attraction to it and to the earth as a whole and there is no lunar tractive force. At these positions the elevation of the water is that of the mean sea level. When the particle of water is at a position where the moon is directly overhead or underfoot (at zenith or nadir), there is also no tractive force and the direction of the tractive force changes. At these positions the cumulative effect of the tractive forces in moving the water toward the moon come to an end and high waters are to be expected.

The response of the water to the tangential tractive forces has the same period as the force in question. Its amplitude and timing, however, vary greatly depending on the character and dimensions of the enclosure in which the water lies. Using a procedure known as *harmonic analysis*, which determines the amplitude and epoch of a constituent of any given period in an observed tide, it has been found that the tide at any position may be described closely by a combination of *partial tides* each having a period corresponding to some motion of the moon or sun relative to the earth. The principal constituents are:

	Period Hours	Speed Number Degrees per Hour
M_2, the lunar semidiurnal constituent	12.42	28.98
S_2, the solar semidiurnal constituent	12.00	30.00
K_1, a lunar diurnal declinational constituent	23.94	15.04
O_1, a lunar diurnal declinational constituent	25.82	13.94
P_1, the solar diurnal declinational constituent	24.06	14.96

There are many additional constituents which may be deduced to result from the astronomical motions. The amplitudes of these constituents are so small that they may be neglected for present purposes.

When the tidal wave enters shallow water, i.e., that in which the depth is not much greater than its amplitude, higher harmonics of the astronomical constituents are developed as discussed in Chapter VI. These harmonics are also considered to be constituents of the tide.

The tide predicting machine used by the Coast and Geodetic Survey in preparing the tide tables actually takes account of 37 constituents. The tide as observed at any position is considered to be due to a spectrum of partial tides each having the form of a progressive wave.

The tide at any position differs in amplitude from day to day, an effect known as the *diurnal variation*. Near the times of new and full moon the diurnal constituents and the solar semidiurnal constituent tend to reinforce the lunar semidiurnal constituent thus producing exceptionally high tides. Such tides are known as *spring tides*. In New England waters the range of spring tides may be as much as 30 percent greater than that of the mean tide. Near the time when the moon is in quadrature the tide is less than the mean tide. Such tides are known as *neap tides*.

While the full spectrum of partial tides is taken into account in developing precise predictions of the range and timing of the tide as it occurs from day to day, this procedure leads to unnecessary complication in showing how and why the tide differs from place to place which is the objective of this study. This is adequately indicated by an examination of the mean tide, which eliminates the day-to-day variation. The *mean tide* recorded in the tide and current tables is obtained by averaging the hourly observations made in the course of a lunar cycle of 29 days or its multiple. This procedure eliminates the effects of the diurnal constituents and of the semidiurnal solar constituent. The mean tide corresponds closely to the partial tide due to the M_2 constituent which is the principal cause of the tide in the North Atlantic Ocean. While the full spectrum of constituents must be used to obtain predictions of the daily variation of the tide at any position,

The Origin of the Tide

such as may be found in the tide and current tables, the mean tide indicates the character of the tide at any place and its variation from other places.

It is assumed that the tide on the continental shelf and at positions within it originates solely from the tide in the deep ocean, not from the direct action of the gravitational force of the moon on the local water as the earth rotates. This assumption is made because the shelf is of such limited extent that significant differences in force do not occur on its opposite sides. This is also true of the Great Lakes in which significant astronomical tides do not occur.

The tides of different bodies of water have been classified according to whether the diurnal constituents or the semidiurnal constituents dominate their behavior, as diurnal, mixed-mainly diurnal, mixed-mainly semidiurnal, and semidiurnal. In the North Atlantic Ocean the tide is usually mixed-mainly semidiurnal being principally due to the M_2 constituent. Such a tide is illustrated by that at Boston shown in Figure I-1. The suppression of the semidiurnal tide by interference in Vineyard and Nantucket sounds causes the diurnal constituents to become relatively more effective so that the tide at Woods Hole is mixed but mainly diurnal (Figure I-2). In Bournes Pond, a small tributary of Nantucket Sound east of Falmouth Harbor, where the very shallow inlet favors the entrance of the spring tides, it becomes nearly diurnal in character at that time (Figure I-3).

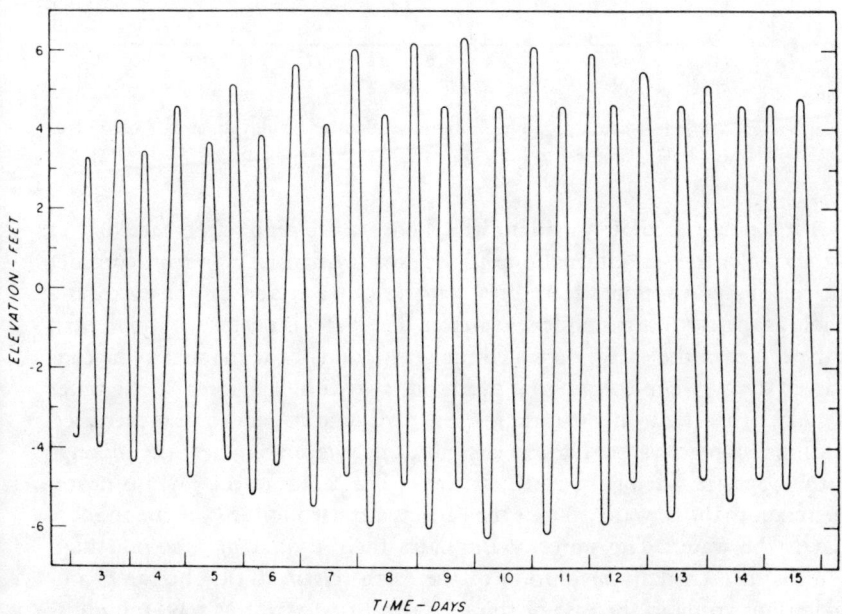

Fig. I-1. Mixed Tide — mainly semidiurnal at Boston. As predicted June 3 – June 15, 1952.

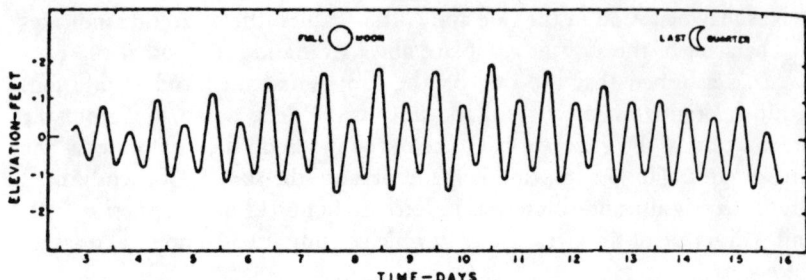

Fig. I-2. Mixed Tide — mainly diurnal at Woods Hole. As recorded June 3 – June 16, 1952. (From the *Journal of Marine Research*, Vol. 12. With permission.)

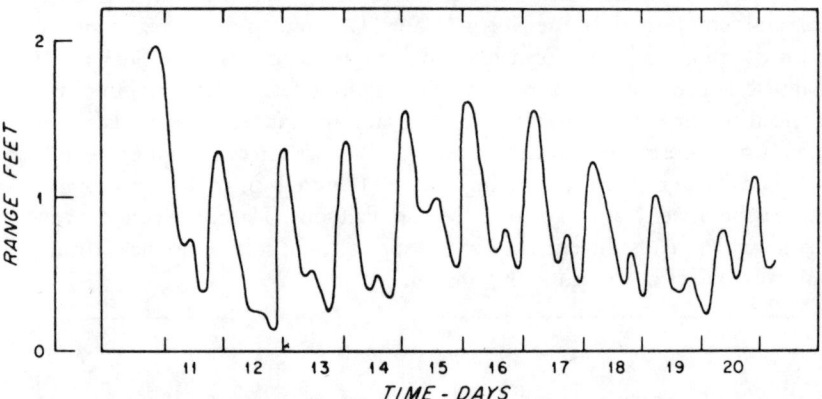

Fig. I-3. Mixed Tide — almost completely diurnal at Bournes Pond, Falmouth. As recorded May 11 – May 20, 1972. (Courtesy of John Moody.)

It is not immediately evident why there are two high waters each day, as is the case of the semidiurnal tide in the North Atlantic. However, consider the relative gravitational force on a particle of water and on the earth as a whole as the earth turns under the moon. The water is nearer the moon than is the center of the earth, on which the gravitation of the moon may be considered to act, when the moon is overhead, at zenith. The force on the water is greater than that on the center of the earth and its elevation is increased leading to a high water. These relative forces are reversed where the moon is at the opposite side of the earth, at nadir. The center of the earth is nearer the moon than the water so that the earth is attracted toward the moon more that is the water. The water accumulates there producing a second high water with each daily revolution of the earth (Figure I-4). The same considerations apply in the case of the solar constituent. In this way, semidiurnal tides may be explained.

The Origin of the Tide

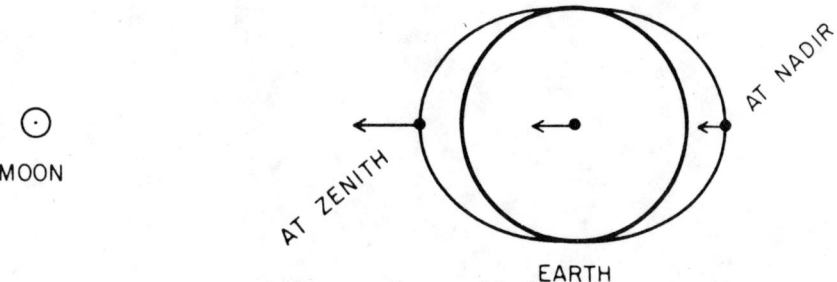

Fig. 1-4. Diagram of production of semidiurnal tide. Length of arrows indicate relative force of moon on earth and on water when moon is at zenith and nadir.

References

Doodson, A. T. and H. D. Warburg. Admiralty Manual of Tides. Her Majesty's Stationary Office, London, 270 pp., 1941.

Schureman, P. Manual of Harmonic Analysis and Prediction of Tides. Government Printing Office, Washington, 317 pp., 1941. Coast and Geodetic Survey Special Publication No. 98. Revised (1940) Edition.

von Arx, W. S. An Introduction to Physical Oceanography. Chapter 3, Tides and Other Waves. Addison-Wesley Publishing Co., Reading, Mass., 422 pp., 1962.

Chapter II

THE PROGRESSIVE WAVE

The progressive wave is a stable form of motion by which energy imparted at a point on the sea surface spreads to other positions. It is that from which the shape of the tidal wave may be derived when two or more progressive waves are present at the same time at any position. The constituents of the tide are considered to have the form of progressive waves in water in which the depth is great relative to the amplitude of the waves.

Characteristics of a Single Progressive Wave

Figure II-1 shows the form of a progressive wave and indicates the terms used in its description. The *period* is the time (in hours) between the passage of successive crests at any position. The *wave length* is the distance traversed by the crest in one period. The *elevation* is the height of the surface relative to mean sea level. The *amplitude* is the height of the surface above mean sea

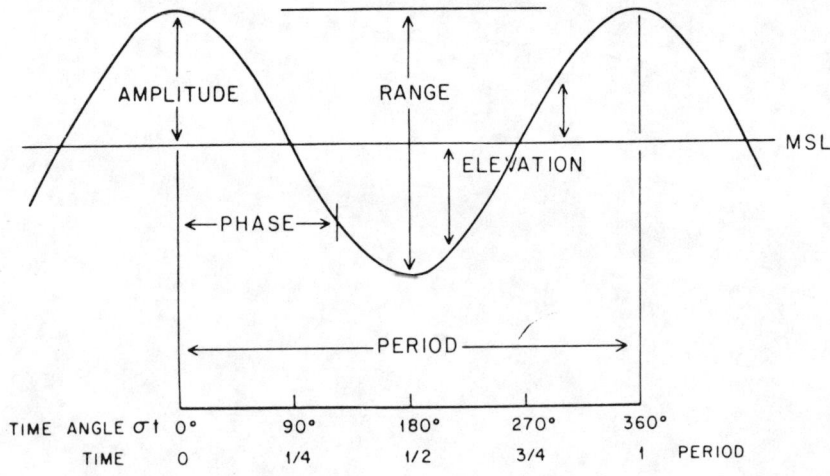

Fig. II-1. The progressive wave.

The Progressive Wave

level when the wave is at its maximum elevation. The *range* is twice the amplitude. The *phase* is the condition of the wave at any time and place. It is the same as the time since high water when they are both expressed in angular notation.

In computations of the theoretical characteristics of the tide an *angular notation* is used which is appropriate for the description of recurring phenomena, such as the tidal elevation, in which positive values alternate with similar negative values when expressed relative to mean sea level. In Figure II-2 the angular notation is compared with the more familiar linear notation. The circumference of a circle, 360°, represents one period of the wave. The origin of time is taken when it is high water at some position, when time is 0°. The radius of the circle represents the amplitude of the wave. The time of any event is represented by a time angle, σt, and the elevation at that time by the elevation of the point where the time angle cuts the circumference relative to the undisturbed elevation of mean sea level (at 90° and 270°). The angular notation permits the time of events to be expressed in degrees and thus allows the application of the laws of trigonometry. Thus the elevation at any time is proportional to the cosine of the time angle σt. The phase of the progressive wave at any position relative to its phase at the origin of time is kx where k is one period of 360° and x is a fraction of the period of the wave.

If the elevation of the water is to increase or decrease at any position, currents must flow to or from that position from elsewhere. In a progressive wave maximal tidal currents flow in the direction in which the wave is moving when its elevation is maximal. Maximal currents flow in the direction opposite to that of its progression when the elevation is minimal. Midway between these two extremes, when the elevation is that of mean sea level, the velocity of the tidal current is zero and it is slack water (Figure II-3).

The rate of advance of a progressive wave in a uniform channel, its *celerity* or *phase velocity*, depends on the depth. In water which is shallow relative to the length of the wave the celerity, C, is given by $C = \sqrt{gd}$ where g is the gravitational constant and d is the depth of water. The celerity of the wave varies with the square root of the depth. Since g equals 32.2 feet per second per second, when the depth is 10,000 feet, as in much of the deep ocean, the rate of progression is 336 knots while if the depth is 100 feet, as in coastal waters, it is reduced to 33.6 knots. Thus a 100-fold decrease in depth produces a 10-fold decrease in the rate of progression. Still greater decreases occur in the shallower waters of coastal embayments and straits.

In natural channels the depth and consequently the celerity of the waves is not constant. It is assumed that under such conditions the form of the progressive wave is unchanged; i.e., the relations between elevation, time, and phase are the same as though the wave were moving in a uniform channel.

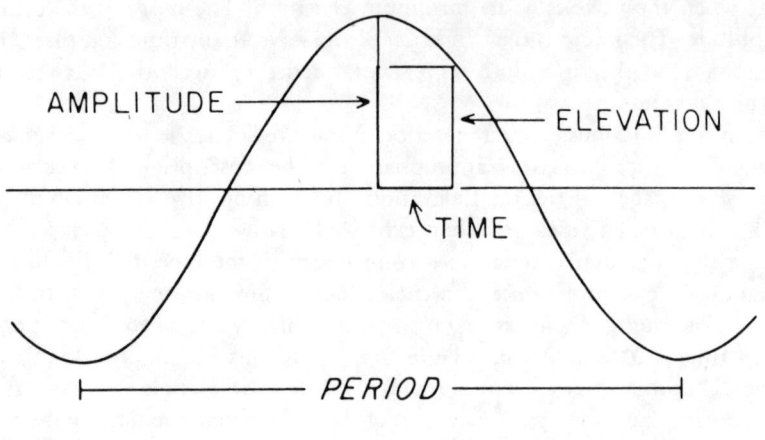

LINEAR NOTATION

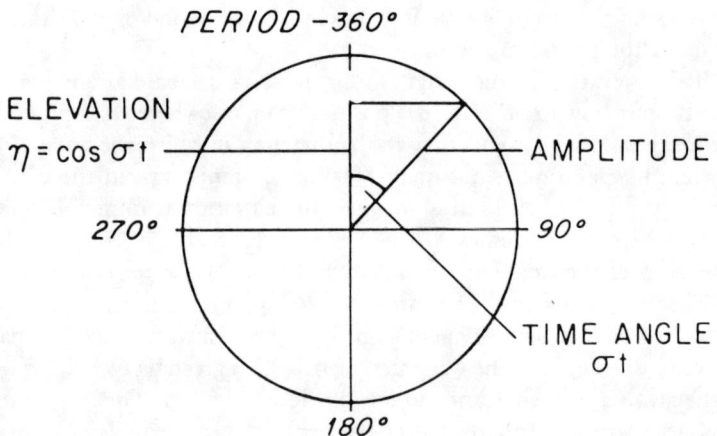

ANGULAR NOTATION

Fig. II-2. Diagrams of the properties of progressive waves in linear and angular notation.

When a progressive wave meets a barrier such as a coastline or a dam, a *reflective wave* is generated moving away from the barrier. When its advance is stopped, the kinetic energy due to its motion is converted into potential energy causing the elevation to increase and to generate a reflected wave moving in the opposite direction. Because the kinetic energy of a pro-

The Progressive Wave

gressive wave is equal to its potential energy, the elevation when it is stopped by the barrier will rise to twice that it has when it is progressing freely. At the position of reflection the entering wave and its reflection are in phase and of equal amplitude. The reflected wave generated at the barrier moves down the channel in a directon opposite to that of the incoming wave and is subject to similar attenuation along its course. It is as though the incoming wave were simply turned back by the barrier.

Attenuation is the decrease in amplitude of a wave in traversing a channel. Progressive waves are decreased in amplitude by friction between the water and the bottom. Reflection occurs from the sides of the channel when they are irregular. Such reflection also decreases the energy of the advancing waves and thus reduces their amplitude. If it is assumed that the attenuation due to these effects is proportional to the change in phase, the decrease in amplitude may be expressed as an attenuation coefficient, μ, which is the change in amplitude per period. Since x is a fraction of a period the attenuation is given by $e^{-\mu x}$.

Where the passage is very irregular in outline, a large part of the attenuation is due to reflection from its sides and the attenuation is considerable.

The Combined Effect of Progressive Waves

When two waves of the same period are moving in opposite directions along a channel, their effects on the tide are additive. This is known as *interference*. If the passage is sufficiently long, there will be a series of positions, separated in phase by 180°, in which the two waves are in phase and are known as *antinodes*. On the New England coast none of the passages are long enough to develop more than one antinode. The position of phase equality may be taken as the origin of phase ($kx = 0$) and the time when it is high water there as the origin of time ($\sigma t_H = 0$). Both waves are at their maximum elevation at this position and time and since they are additive the maximum elevation and range of the tide will occur there. At other positions the two waves are not in phase. The greatest difference in elevation will occur where the two waves will differ in phase by 180°, as where $kx = -90°$. At this position a *node* will occur at which the interference is maximal and the elevation and range well be the smallest to occur in the passage.

If the passage is not sufficiently long, the position of phase equality may be virtual inasmuch as it does not occur within the passage. The elevation and time of high water there may be obtained only by extrapolation of the theoretical equations. Many straits and embayments are so short that nodes do not occur within their length.

When the two waves are of the same amplitude and there is no attenuation, the combined wave which results has the form of a *standing wave*. The surface rises or falls simultaneously on either side of a node and there is no evident progression. Such standing waves may be observed where wind waves are reflected from a vertical sea wall. The currents which produce the changes in elevation of a standing wave are quite different from those of single progressive waves such as are shown in Figure II-3. At the node when the crest of the wave moving toward the antinode or barrier is passing, the current due to that wave is flowing in the direction of its progression. The other wave, which is 180° out of phase, is moving in the opposite direction but its current is also flowing in a direction opposite to its direction of progression. As a result, the currents due to the two waves flow in the same direction and reinforce one another. Maximum currents occur at the node in the direction in which the water is rising. One-half period later the direction of the currents are reversed and maximum currents occur in the opposite direction.

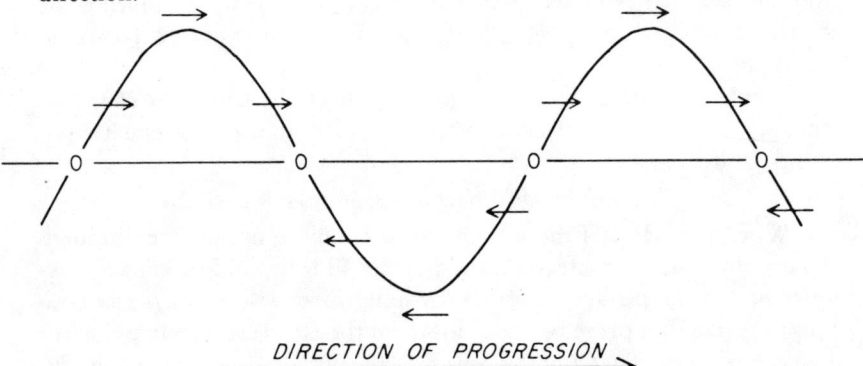

Fig. II-3. Currents in a progressive wave. Arrows show direction of horizontal flow and roughly its velocity in relation to elevation.

In natural channels there is always some attenuation so that in the case of embayments the entering wave is larger that the other except at the position of reflection. The difference appears as a progressive wave superimposed on the standing wave. Its result is to cause the high water interval to increase in passing toward the position of reflection, an observation which is good evidence that the tidal waves are attenuated in these channels.

Theory

Combining the ideas that the elevation is proportional to the cosine of the changes in phase due to time and position and that the wave undergoes an attenuation given by $e^{-\mu x}$, the elevation at any time and place, η_1, of the progressive wave which moves in the *x direction in a channel* may be written:

$$\eta_1 = A \cos(\sigma t - kx)e^{-\mu x}.$$

The Progressive Wave

The elevation, η_2, of the wave moving in the opposite direction, in the $-x$ direction, is given by a similar equation in which the sign in the terms where x occurs is changed.

$$\eta_2 = RA \cos(\sigma t + kx)e^{\mu x}.$$

In these equations A is the amplitude at the origin of time of the wave moving in the x direction. R is the ratio at the origin of time of the amplitudes of the wave moving in the $-x$ direction to that of the wave moving in the x direction, σt and kx are changes in phase due to time and place respectively, and x is a fraction of the period of 360°. The tide resulting from the simultaneous presence of two progressive waves moving in opposite directions is due to their interference. At any time and place its elevation, η, is given by the sum of these waves:

$$\eta = \eta_1 + \eta_2 = A \cos(\sigma t - kx)e^{-\mu x} + RA \cos(\sigma t + kx)e^{\mu x}.$$

Equations may be derived from this expression which give the elevation at high water and the intervals of high and slack water as a funciton of kx, the phase at the position. These are:

The elevation at high water, η_H, is given by

$$\eta_H = A \frac{(\cos kx - F \tan kx \sin kx)e^{-\mu x} + R(\cos kx + F \tan kx \sin kx)e^{\mu x}}{1 + F^2 \tan^2 kx}$$

where

$$F = \frac{Re^{\mu x} - e^{-\mu x}}{Re^{\mu x} + e^{-\mu x}}.$$

High water occurs when the time angle, σt_H, is given by

$$\sigma t_H = \tan^{-1}(-\tan kx \cdot F).$$

Slack water (ebb begins) occurs when the time angle, σt_s, is given by

$$\sigma t_s = \tan^{-1} \frac{F}{\tan kx}.$$

The values of R and μ characteristic of a given passage and of kx characteristic of positions in that passage may be obtained by nomographic analysis as described in the following chapter. When these values are introduced into the theoretical equations, the results may be compared with the predictions given in the tide and current tables as a test of the validity of these equations.

In Appendix A some simple observations and experiments are described which should make the phenomena mentioned in this chapter more real to the reader.

References

Doodson, A. T. and M. D. Warburg. Admiralty Manual of Tides. Her Majesty's Stationary Office, London, 270 pp., 1941.

Redfield, A. C. The analysis of tidal phenomena in narrow embayments. Papers in Physical Oceanography and Meteorology. Vol. XI, No. 4, 36 pp., 1950.

Redfield, A. C. Interference phenomena in the tides of the Woods Hole region. Journal of Marine Research, 12, 121-140, 1953.

Redfield, A.C. The tide in coastal waters. Journal of Marine Research, 36, 255-294, 1978.

Chapter III

NOMOGRAPHIC ANALYSIS - THE TIDE IN STRAITS

There are two modes of motion which account for the tide at many of the positions on the New England coast and which depend on the interference of waves moving in opposite directions. They may be described by the theoretical equations given in the preceding chapter. These are in straits in which the interference is between waves entering the passage at opposite ends and in embayments in which the interference is between the entering wave and its reflection from the head of the embayment. A third mode of motion is due to the entering wave alone, apparently not influenced significantly by waves reflected from upstream. It is unusual but occurs in the lower reach of the Hudson River and is considered in Chapter XIII.

In the following discussion it is assumed that these passages are freely open to the outer sea, that higher harmonics are absent, that they are so narrow that the tide is not influenced by the rotation of the earth, that the weather may be neglected, and that the mean sea level remains constant. Modifications by these factors are considered in Chapters V to IX.

Nomographic Analysis

The theoretical equations give properties of the tide as a function of the phase, kx, at positions along the channel. The tide and current tables do not give the value of kx at the positions listed but these may be determined by nomographic analysis. Nomograms are prepared so that the data listed in the tide tables, after suitable conversion, may be entered directly in the diagram. The time angle of high water σt_H at given values of kx is entered as the abscissa and the logarithm of the ratio of the amplitude at these values of kx, η_H, to that at the position of phase equality, η_{HO}, i.e., log η_H / η_{HO} is entered as the ordinate. Since both R and μ may vary, it is convenient to prepare separate nomograms for each of a series of R in which lines represent the values of log η_H / η_{HO} and σt_H at various values of μ. A second series of lines may be drawn, each line representing the corresponding value of kx. Examples of such nomograms are shown in Figure III-1.

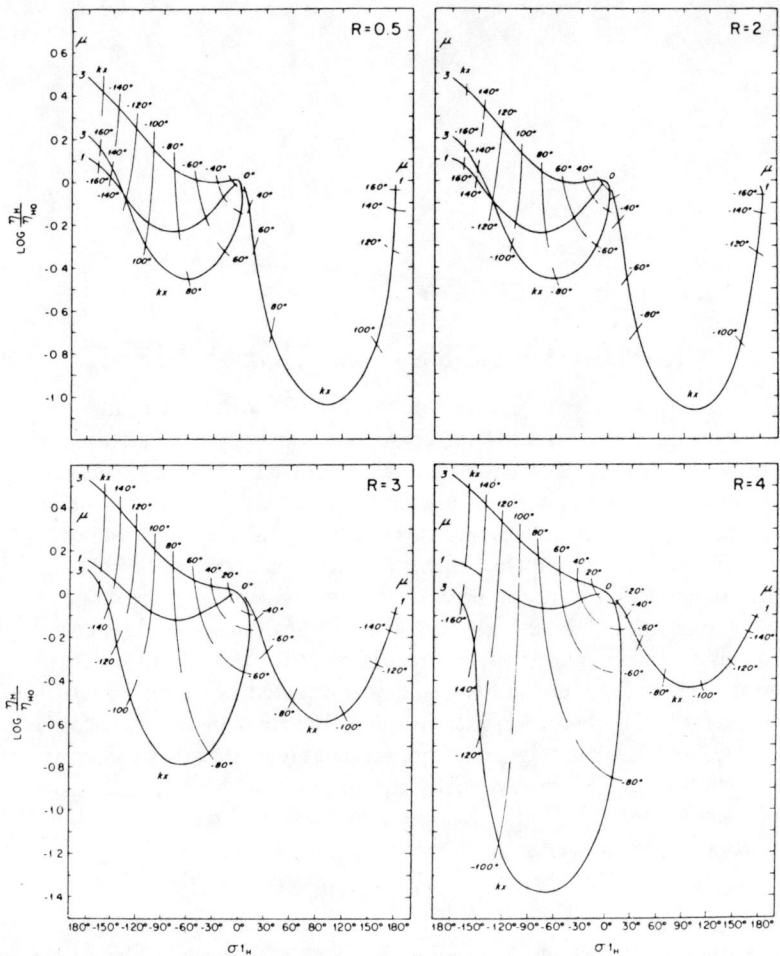

Fig. III-1. Nomograms for passages in which $R = 0.5$, 2, 3, and 4 and in each of which $\mu = 1$ and 3. For nomograms when $R = 1$ and $R = 0$ see Figure IV-2 and XIII-2 respectively. (From the *Journal of Marine Research*, Vol. 36. With permission.)

To use these nomograms first find the values of R and μ to be assigned to a particular passage and the values of kx to be assigned to positions along that passage, then the characteristics of the tide in a real passage may be considered to correspond to those of a segment of a theoretical curve in an appropriate nomogram. Before this correspondence can be demonstrated, the time of high water and slack water which are given in hours in the tables must be corrected for harmonics and converted into the angular notation of

Nomographic Analysis—The Tide In Straits

the theoretical equations. The time of events is obtained by adding to the Greenwich intervals of the reference stations, shown in Table I, the time difference for the particular positions given in the tables. The times so obtained are corrected for the presence of harmonics as discussed in Chapter VI. The result is in hours. It is converted to the angular notation by multiplying by 29° per hour.

The characteristics of each position are entered on a grid similar to that used in the nomograms as points of which the ordinate is the range plotted as its logarithm and the abscissa is the Greenwich high water interval in degrees. A line can usually be drawn through these points representing the relation of the logarithm of the range to the interval of high water. Because the range is entered as its logarithm, the shape of this line is independent of the magnitude of the range. Adjustments may be applied for the relation of $\log R$ to $\log \eta_H / \eta_{HO}$ and of the high water interval to σt_H which will bring this line into correspondence with *some segment* of a line on an appropriate nomogram. Since at the position of phase equality σt_H and $\log \eta_H / \eta_{HO}$ are both equal to zero, the values of these adjustments are given by the logarithm of the mean range and the high water interval at that position. If this can be done, the values of R and μ given by the nomogram can be applied to the passage in question and those of kx to positions along that passage (Figure III-2).

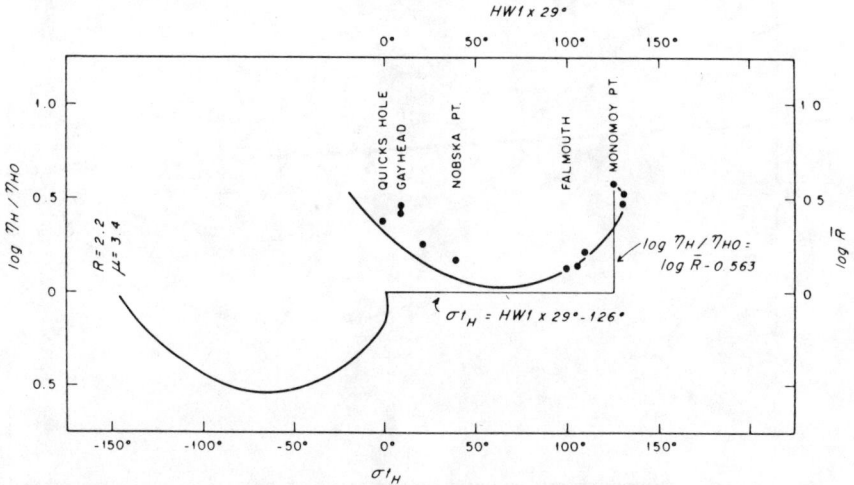

Fig. III-2. Adjustment of data for predicted range and high water interval to the selected theoretical curve. At right the predictions (corrected for harmonics) are entered. The curve drawn through these points is identical in shape with that at left which is calculated for $R = 2.2$ and $\mu = 3.4$. Monomoy Point is taken as the position of phase equality. The adjustments are given by the transposition of this point to the position of phase equality of the theoretical curve.

The method of nomographic analysis is somewhat subjective but the validity of its results may be tested objectively by comparing, as a function of kx, the range and times of high and slack water given by the tables with those obtained when the values of R, μ, and kx given by the nomographic analysis are entered into the theoretical equations. These values are found to agree sufficiently to differentiate between the properties of most of the passages to which the analysis has been applied.

The Tide in Straits

The theoretical equations given in Chapter II apply to the tides in straits. R may have any value depending on the relative magnitude of the waves which enter at either end of the passage. In the waters of New England and New York straits occur in which R varies from 0.2 to 2 and μ from 1.2 to 7.

Vineyard and Nantucket sounds together form a strait which serves as an example. This strait connects the water of the continental shelf off Gay Head with that of the Gulf of Maine off Monomoy Point. There is also a connection with the continental shelf through the Muskeget Channel but as

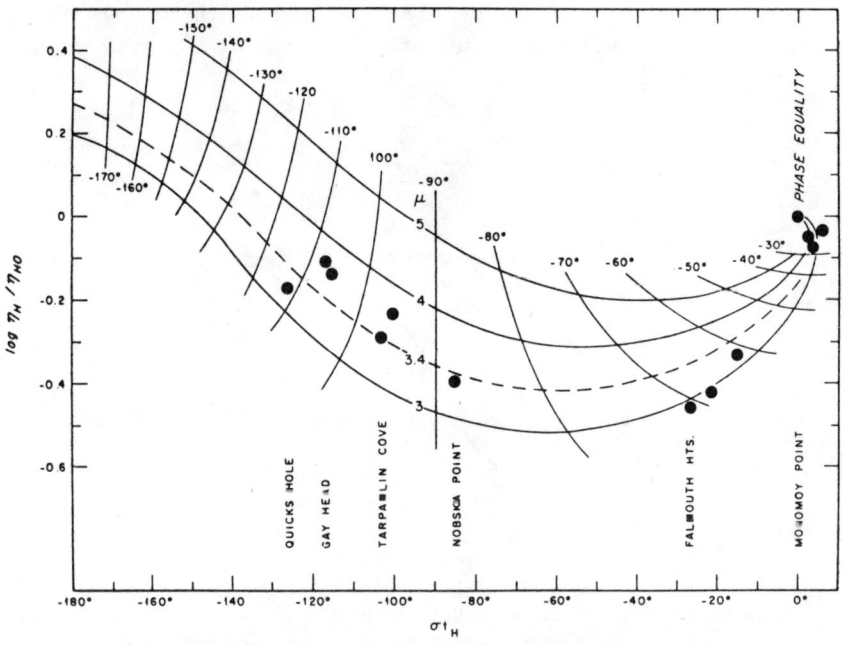

Fig. III-3. Nomogram for a passage in which $R = 2.2$ on which are entered points representing positions on the Vineyard-Nantucket Sound system after adjustment and correction for harmonics. The values of kx assigned to these positions may be read approximately from the nomogram.

Nomographic Analysis—The Tide In Straits

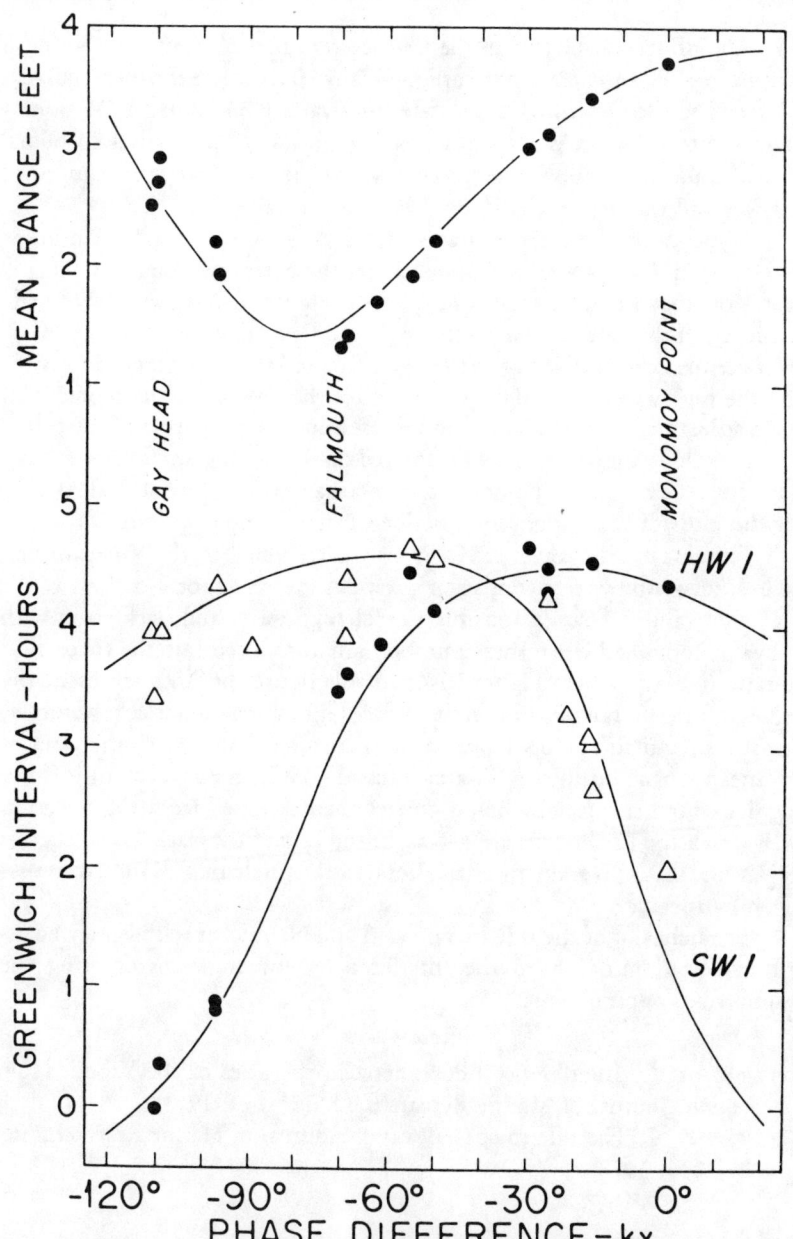

Fig. III-4. The Vineyard-Nantucket Sound system. Comparison as a function of kx of the theoretical values of mean range and intervals of high water (HWI) and slack water (SWI) which are shown as solid lines in the case of the theoretical relations and as points based on the predictions. (From the *Journal of Marine Research*, Vol. 36. With permission.)

this only influences the tide in the southeastern part of Nantucket Sound it is neglected. High water occurs earliest at Gay Head where the mean range is 2.8 feet and the Greenwich high water interval is 0.34 hours. At Monomoy Point the mean range is 3.7 feet and the high water interval is 4.33 hours. Near Falmouth Heights a node occurs where the mean range is reduced to 1.3 feet and the high water interval is 3.45 hours.

Nomographic analysis indicates that for the Vineyard and Nantucket sound system $R = 2.2$ and $\mu = 3.4$ and that the position of phase equality is near Monomoy Point where $A = 0.578$ foot. Figure III-2 shows how the adjustments of the data for the mean ranges predicted and the time interval of their occurence are arrived at. After the predicted values for the mean range and the high water interval are corrected for the presence of harmonics and these adjustments are made, points representing the properties of the tide at positions along the system may be entered on the nomogram for $R = 2.2$ as is done in Figue III-3. The points scatter about a line for $\mu = 3.4$. Values of kx for the individual positions are obtained from the nomogram.

The test of the values of A, R, and μ assigned to the Vineyard and Nantucket Sound system consists of a comparison, as a function of kx, of the predicted values of range and intervals of high water and slack water with the values obtained when these numbers are introduced into the theoretical equations as is done in Figure III-4. In this figure the lines represent the values of mean ranges and intervals of high water and slack water, as calculated, and the points represent their predicted values. In the cases of the mean range and the high water interval (HWI), the agreement is close, the departures not usually being greater than expected from the precision with which the predictions are given. In the case of the slack water interval (SWI), the scatter is greater but the trend of the predictions is similar to that given by theory.

The behavior of the tide in Vineyard and Nantucket sounds may be attributed predominantly to the interference between waves entering the system from opposite ends.

References

Redfield, A. C. Interference phenomena in the tides of the Woods Hole region. Journal of Marine Research, 12, 121-140, 1953.

Redfield, A. C. The tide in coastal waters. Journal of Marine Research, 36, 255-294, 1978.

Chapter IV

THE REFLECTED CO-OSCILLATION OF EMBAYMENTS

The tide in an embayment is an oscillation due to the interference of a progressive wave generated at its mouth and a wave moving in the opposite direction formend by its reflection from the head of the embayment.

At the position of reflection, the entering wave and its reflection are always of the same phase and, assuming reflection to be complete, of the same elevation. R is consequently always equal to 1. The tide in embayments may be considered to be a special case of the general theory on which the theoretical equations given in Chapter II are based and may be represented by a single nomogram for $R = 1$. Mathematically one cannot distinguish between the behavior of the tide in such a passage below the antinode and in an embayment in which it is assumed that reflection takes place at this position.

In a progressive wave the kinetic energy due to its motion is equal to the potential energy due to its elevation. At the position of reflection the motion is stopped and the kinetic energy is converted into potential energy. The elevation of the water surface at the position of reflection will consequently rise to twice that which the entering wave would have, after allowing for its attenuation in moving along the length of the embayment. One may expect the amplitude or range of the tide to increase in ascending an embayment and to be maximal at the position of reflection unless the attenuation is great.

Augmentation may be defined as the increase in the amplitude of the tide in ascending an embayment. It depends upon interference between the entering wave and its reflection and consequently on the length of the passage and is decreased by attenuation. Figure IV-1 shows the degree in which the range of tide is augmented at the position of reflection relative to that at the entrance. If the attenuation is small, the augmentation is maximal when the phase change within the channel is about 90° (or ¼ of the wave length). As the attenuation becomes greater, the augmentation is less and its maximum occurs when the phase changes within the channel become less than 90°. If the attenuation coefficient is nearly 4 or greater, there is no augmentation.

The maximum augmentation of the tide in certain embayments has been explained as the result of *resonance* which occurs when the natural period of oscillation of the embayment corresponds to ¼ period of the tidal wave. If this were so and if there were no attenuation, resonance would be additive so that the range at the position of reflection would be infinite. In natural embayments there is always some attenuation so that the maximum augmentation is limited as shown in Figure IV-1. The concept of resonance has not been helpful in explaining the tides of waters of New England and New York which are accounted for adequately by the theoretical equations given in Chapter II and developed from the idea that they result from the interference of waves moving in opposite directions.

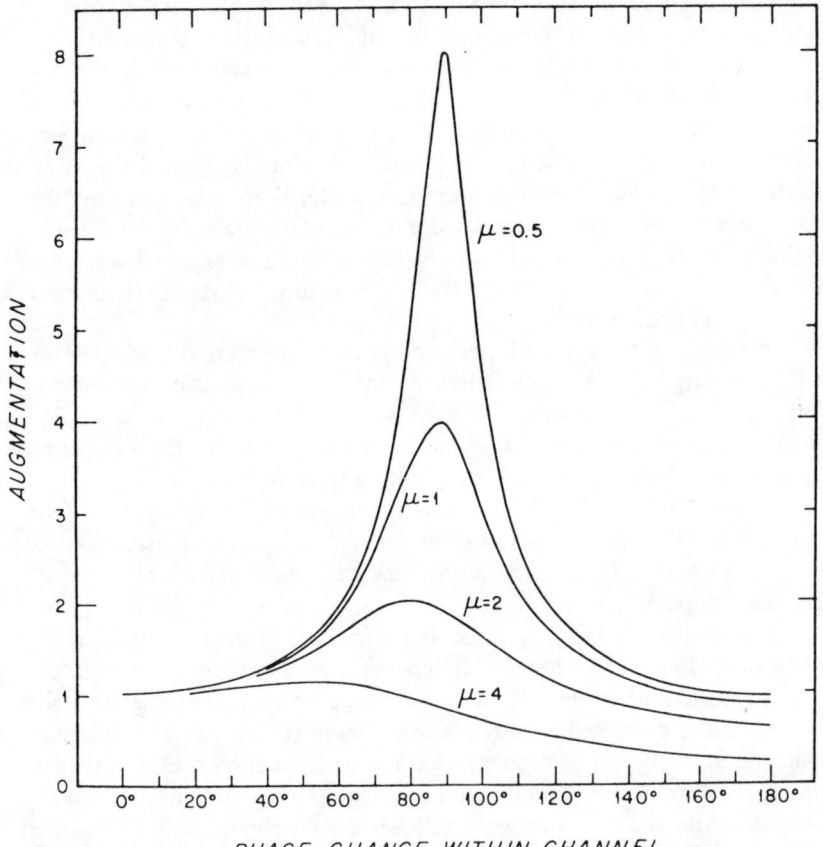

Fig. IV-1. Augmentation of tide in embayments as a function of the phase change within the channel and the value of the attenuation coefficient, μ.

The Reflected Co-oscillation of Embayments

If there were no attenuation, the form of the tidal wave would be a standing wave rising and falling simultaneously on either side of the node. In natural embayments, where there is always some attenuation, a standing wave occurs at the position of reflection but is modified below this position by a superimposed progressive wave, as discussed on page 10.

Figure IV-2 is the nomogram for a passage in which $R = 1.0$ and applies whether the passage is a strait or an embayment. It shows the theoretical relations of elevation at high water, the time angle of high water and the phase differences between positions along the passage and the position of phase equality (which in the case of embayments is the position of reflection). Note that at positions within the passage the time angle of high water is always negative, indicating that in embayments high water occurs latest at the position of reflection. A *node* occurs where the elevation is smallest. The change in phase (kx) is about 90° if μ is small. As μ increases the position of the node shifts toward the position of reflection.

The velocity of the tidal current is zero at the position of reflection, where the flow is stopped. In a channel of uniform cross section the flow will

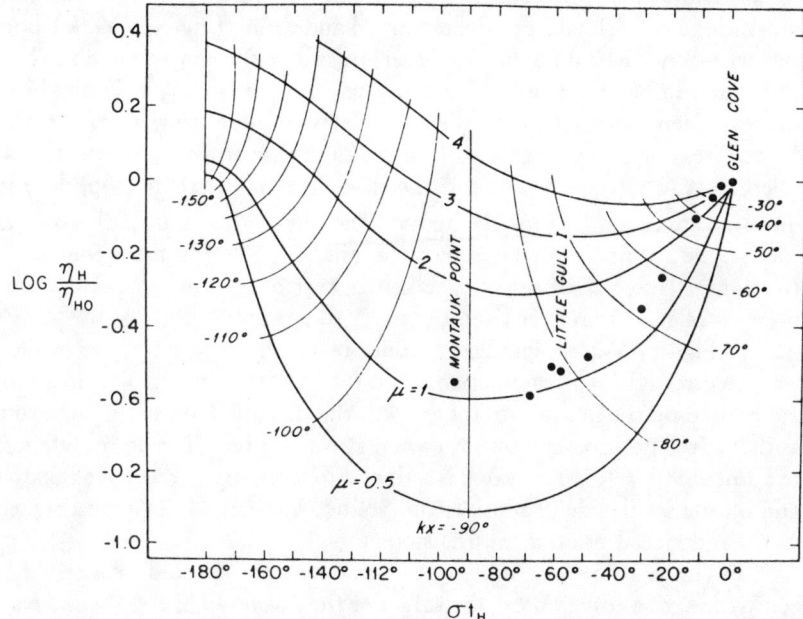

Fig. IV-2. Nomogram for a passage in which $R = 1.0$, which applies to such straits and to all embayments. The points represent the tide at positions in the Long Island Sound system, corrected for harmonics and transformed to the angular notation, taking the tide at Glen Cove to represent that at the position of reflection. (From the *Journal of Marine Research*, Vol. 36. With permission.)

increase toward the node past which all the water which causes the tide to rise or fall within it must flow. In a natural embayment, in which the depth and width vary, no general statement can be made regarding the velocity. However the time of slack water, when the entering wave and its reflection are of the same amplitude at any position, is independent of the dimensions of the channel.

The predicted elevations of range, expressed as log η_H/n_{HO}, and times of high water, properly corrected for harmonics (as discussed in Chapter VI) and transformed into the angular notation, may be entered on the nomogram for $R = 1$. If they fall along a line for a given value of μ, its value may be applied to the embayment in question and values of kx may be assigned to the individual positions along its length. The validity of the values of R, μ, and kx so obtained may be tested by comparing, as a function kx, the values for range and times of high and slack water given when these numbers are entered into the theoretical equations with those predicted in the tables.

The *Long Island Sound system* may be taken as an example of the tide in an embayment. Actually, Long Island Sound is a part of a strait separating Long Island from the mainland and connecting New York Upper Bay with Block Island Sound. However, its tide has been considered to be a classical example of an embayment with reflection occurring at Throgs Neck or near Glen Cove. Such treatment neglects the tidal flow from the East River. Treating the passage as a strait indicates that an antinode occurs near Glen Cove where the flow from the East River is equal, but opposite in direction, to that ascending the Sound. The tide at this antinode behaves as though the antinode was the position of reflection. On this basis, the tide in the Long Island Sound system is taken as that of an embayment with the position of reflection near Glen Cove.

In Figure IV-2 the predicted values of mean range and time of high water, corrected for harmonics and transformed into the angular notation, are plotted on the nomogram for $R = 1.0$, the tide at Glen Cove being taken as that at the position of reflection where $A = 1.825$ feet. The points fall near to a line for $\mu = 1.0$. This value is assigned to the tide in Long Island Sound and on the south side of Block Island Sound. Values of kx may be assigned to the individual positions in the system.

In Figure IV-3 the validity of these values of R, μ, and kx is tested by comparing, as a function kx, the values of the mean range and the intervals of high and slack water when these are introduced into the theoretical equations with the prediction for the individual positions. In the cases of the mean range and the high water interval (HWI), the agreement is as good as may be expected from the precision with which the predictions are given. In the case of the slack water interval (SWI), the predictions scatter from the

The Reflected Co-oscillation of Embayments 23

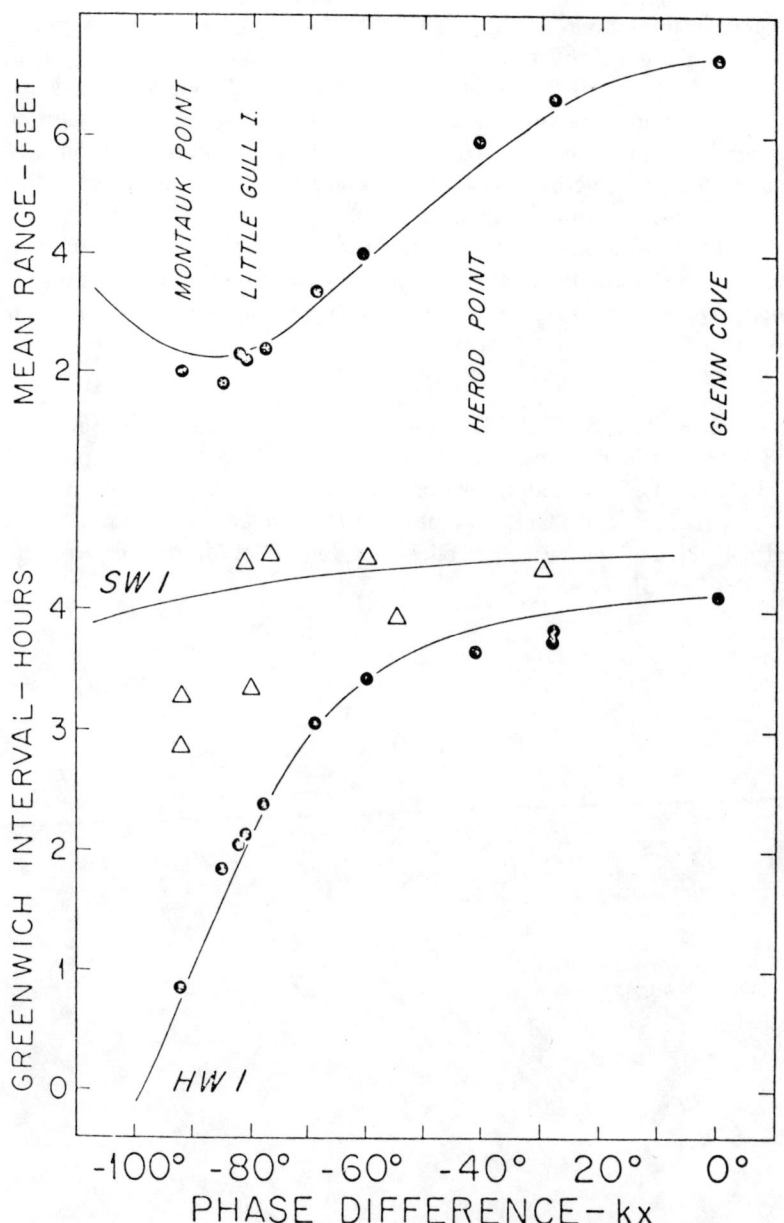

Fig. IV-3. The Long Island Sound system. Comparison as a function of *kx* of the ranges and intervals of high (HWI) and slack water (SWI) calculated from the theoretical equations with points based on the predictions. (From the *Journal of Marine Research*, Vol. 36. With permission.)

theoretical line more widely, and near Montauk Point slack water is predicted to occur as much as one hour earlier than indicated by theory. This may be due to local hydraulic effects in this neighborhood.

Similar nomographic analysis has shown that in other embayments on the New England coast the predicted mean ranges at positions along the channel and the time of high water are also satisfactorily indicated by the theory that they are due to the interference of the entering progressive wave and its reflection from the head of the embayment. In two of these cases, Great South Bay and Peconic Bay, where the channel is constricted at the entrance, hydraulic effects must be assumed there which cause slack water to occur earlier than expected.

References

Doodson, A. T. and D. H. Warburg. The Admiralty manual of tides. Her Magesty's Stationary Office, London, 270 pp., 1941, See Chapter 19.

Redfield, A. C. The analysis of tidal phenomena in narrow embayments. Papers in Physical Oceanography and Meteorology, 11, 1-36, 1950.

Redfield, A. C. The tide in coastal waters. Journal of Marine Research, 36, 255-295, 1978.

Chapter V

HYDRAULIC CURRENTS

In the two preceding chapters the theory of the tide has been developed on the assumption that the passage is freely open to the tide in the outer sea, that higher harmonics are absent, that it is so narrow that it is not influenced by the rotation of the earth, that the weather may be neglected and that the mean sea level is constant. Where these assumptions are not met, the behavior of the tide may be modified, as indicated in this and the following chapters.

Hydraulic currents develop frequently and are often strong in straits where the range and time of high water differ at either end and thus cause a hydraulic head to develop which gives rise to currents additional to those accompanying the tidal waves. Strictly speaking, they are not tidal as they may occur in passages which connect two tideless lakes which have different elevations as in the Niagara River and Falls, to give an extreme example. Their rhythm is due to the tidal rhythm with which the water at either end may rise or fall. Hydraulic currents may also occur in embayments when the mouth is restricted causing differences in elevation to develop between the water in the outer sea and at positions beyond the constriction, or they may develop locally within the embayment wherever the flow is sufficiently retarded.

Hydraulic currents do not disturb the relation between the mean range of the tide and the high water interval as calculated from the interference of the progressive waves moving along the passage in opposite directions. They do, however, greatly affect the time of slack water which usually occurs earlier than calculated from the interference of the progressive waves. If the current were due entirely to the hydraulic effect, slack water would occur at the same time at all positions along the channel, when the elevation at the opposite ends is equal. This is practically true if the hydraulic current dominates the flow through the passage.

The *Harlem River* is an example of a strait in which the flow is due predominatly to a hydraulic current. At the end of the strait where a pro-

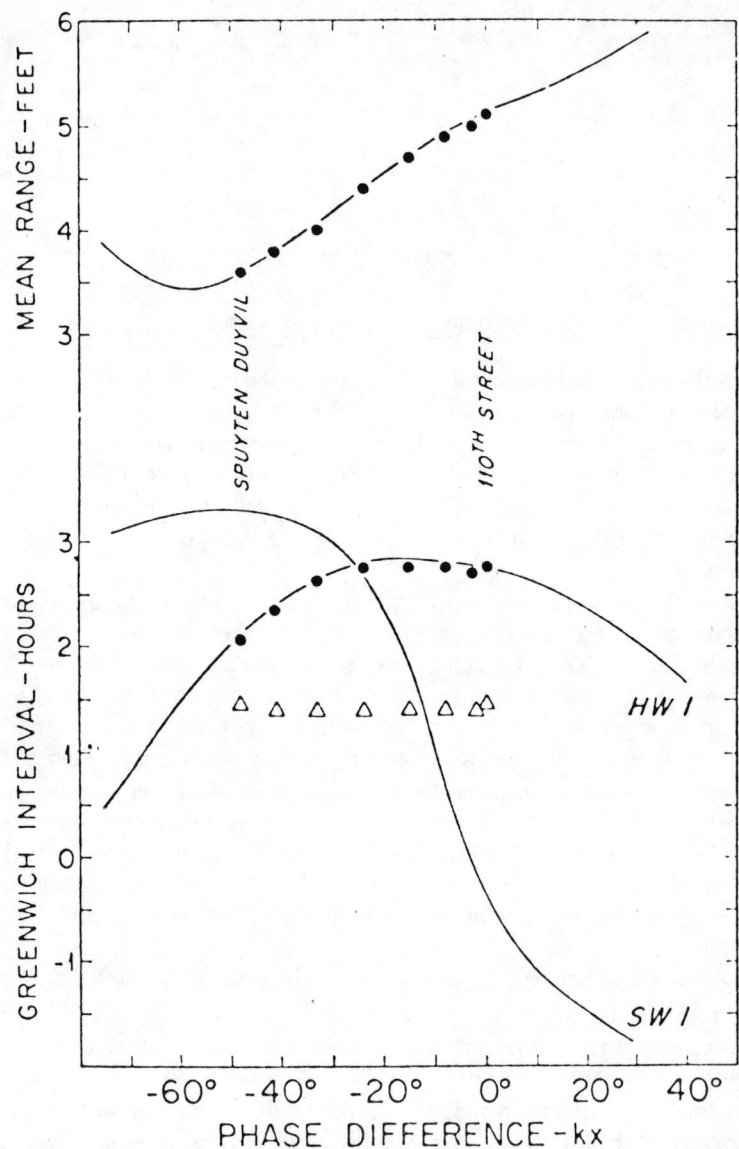

Fig. V-1. Harlem River. Comparison as a function of *kx* of the range and Greenwich intervals of high (HWI) and slack water (SWI) calculated from the theoretical equations with points based on the prediction, when $A = 0.85$ ft., $R = 2$ and $\mu = 5$. Triangles indicate the predicted interval when slack water ebb begins. (From the *Journal of Marine Research*, Vol. 36. With permission.)

gressive wave enters from the Hudson River, at Spuyten Duyvil, the predicted mean range is 3.6 feet and high water occurs 2.07 hours after the moon's transit at Greenwich. At the opposite end, where a progressive wave enters from the East River at 110th Street, the predicted mean range is 5.1 feet and the Greenwich high water interval is 2.77 hours. These differences result in a head between the ends of the strait to which the hydraulic current is due. Nomographic analysis indicates that in Harlem River $R = 2$ and $\mu = 5$, and that the position of phase equality is near 110th Street where $A = 0.85$ foot. Figure V-1 shows the results of the analysis plotted according to the linear notation. The lines represent the values of the mean range and the Greenwich intervals of high (HWI) and slack (SWI), as a function kx, calculated from the theoretical equations when $R = 2$ and $\mu = 5$ and the points represent their predicted values. In the cases of the mean range and the high water interval (HWI), the aggreement is excellent. In the case of the slack water interval (SWI), the predicted intervals are different from those indicated by theory, and slack water is predicted to occur almost simultaneously along the strait. This indicates that hydraulic currents dominate the flow in the Harlem River.

Reversing Falls

Reversing tidal falls may be considered to be extreme cases of hydraulic currents. The most striking example, indeed the most spectacular of tidal phenomena occuring on the New England coast, is the *Reversing Falls of the St. John River* (Figure V-2). The river widens near its mouth, then passes through a narrow and shallow gorge to enter St. John Harbor. In St. John Harbor the mean range of the tide is about 21 feet, high water occuring about 3.35 hours after the moon's transit at Greenwich. Above the gorge (at Indian Town) the mean range is only 1.2 feet and the Greenwich high water interval is 4.8 hours. These differences, particularly in the range, cause a large head to develop in the gorge which gives rise to the strong currents which occur over the sill at that position. The relations are shown diagrammatically in Figure V-3. Note that the level in the river is about 15 feet above low water in St. John Harbor and that, as a result, the head and flow are much greater during the ebb when the water is flowing out of the river than when it is flowing in. This is due to the natural flow of the river. The water in flowing across the sill does not fall as the name suggests. Rather, it flows as a very turbulent rapid. The current turns when the elevation in St. John Harbor is about 15 feet above low water and is equal to that of the river. At those times there is a short period when the current is so small that the passage may be safely navigated.

A somewhat similar but less spectacular situation occurs at Sullivan, Maine, and is known as *The Tidal Falls*. A narrow strip of land juts into Sullivan Harbor reducing its width about three-quarters. A channel not

Fig. V-2. Reversing Falls of St. John River during the ebb.

more that 10 feet in depth connects the upper and lower portions of Sullivan Harbor. As a result of this constriction, the mean range of tide which is 10.5 feet below the constriction is reduced to 6.5 feet above it, while the time of high water slack is delayed about 1.3 hours and of low water slack about 1.75 hours relative to that below the constriction. As a result, heads develop across the constriction which cause strong reversing currents. The coast pilot warns that these currents are swift and dangerous and that navigation through the falls is only safe at slack water.

Tide Mills

The exploitation of the tide as a source of energy or power depends on the creation of hydraulic currents by the head developed between high and low tide. Its practicality depends on a location such that the necessary dams and mill races can be constructed economically. The potential energy of the tide depends on the square of the difference in elevation at high and low water; that is, on the square of the range. If the range is doubled, say from 5 to 10 feet, the energy is increased four-fold. The power which may be attained by a tide mill thus increases rapidly as the range in tide increases.

Tide mills were built where a cove or salt marsh creek could be dammed so as to make a basin in which the water could be kept near the high tide

Hydraulic Currents

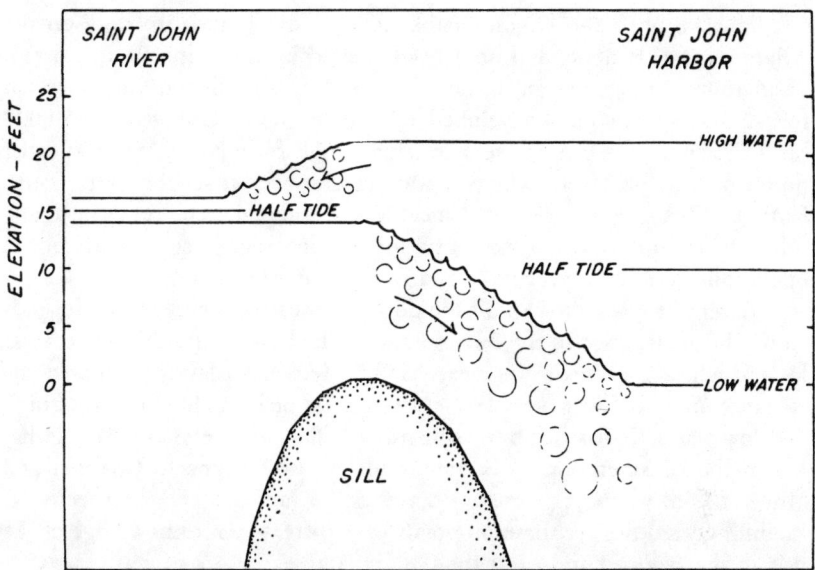

Fig. V-3. Diagram showing elevations of water across the Reversing Falls of the St. John River at high water, half tide and low water in St. John Harbor.

level, in which case they could be operated during about 6 hours of the lower stages of each tide. Where a second basin could be constructed to be kept near the low tide level, they could be operated continuously.

In the early years of settlement and extending well into the 19th century the coastal communities were small and tide mills served to grind the grain and saw the wood for the local people. A tide mill was built in Brooklyn, New York, as early as 1636 and a number of tide mills were operated later at the western end of Long Island Sound where the mean range is about 7 feet. Tide mills appear to have been scarce on the southern shore of New England where the mean range does not generally exceed 4 or 5 feet. North of Cape Cod, where the mean tide range is between 9 and 10 feet, tide mills were quite common. One existed at Barnstable, Massachusetts, made by damming a salt marsh creek (Maraspin Creek) to create a basin within which water could be impounded at the high water level. At Boston tide mills were a principal source of energy. On the coast of Maine the irregular topography of the coast provided many situations where tide mills could be built. One at Kennebunkport is still intact and is used as a restaurant. Another at East Boothbay was intact and presumably operative in 1942 but has since been demolished to make way for highway construction and is marked only by the spillway which passes under the road.

A tide mill at South Harpswell, on Casco Bay, was built in 1867 and is said to have been the largest in the state of Maine and probably on the Atlantic coast. It depended on a head created by damming Basin Cove, a basin about 2 miles in length, near its mouth. This mill could be operated for 12 hours each day, developed 6000 horsepower and ground 50,000 bushels annually. Corn was sent in large vessels from New York and other ports to Portland, from where it was transhipped in schooners to South Harpswell because water-ground meal was considered to be superior to any other. With improved rail transportation and increasing use of steam power operations became unprofitable and came to an end in 1885.

Recent proposals to exploit the tides as a source of energy have depended on the great range in the Bay of Fundy. The Passamaquoddy Project was located where the mean range exceeded 20 feet and islands facilitated the construction of basins where the elevation could be kept close to that of high and low water. It has not been constructed. Recent interest in Canada has led to the consideration of barriers built in the Chegnecto Channel and Minas Basin where the mean range is 33 and 38 feet respectively. Preliminary studies for these proposals have increased our knowledge of the tide in the Bay of Fundy and the Gulf of Maine.

References

Redfield, A. C. The tide in coastal waters. Journal of Marine Research, 36, 255-294, 1978.

Hydrographic Office, U. S. Navy. Nova Scotia Pilot. U. S. Government Printing Office, Washington, 1930. See page 102 on St. John River.

National Ocean Survey. United States Coast Pilot I. Atlantic Coast Eastport to Cape Cod. Washington, 1971. See page 95 on Sullivan Falls.

Zincles, M. and M. Early American Mills. Bramhill House, New York, no date.

Meigs, P. Energy in early Boston. New England Historical and Geneological Register, 128, 83-90, 1974.

Thomas, M. S. South Harpswell's old tide mills. Down East, 19, 21-23, 1972.

Duff, G. F. D. Tidal resonance and tidal barriers in the Bay of Fundy system. Journal of the Fisheries Research Board of Canada, 27, 1701-1728, 1963.

Chapter VI

SHALLOW WATER TIDES AND HARMONICS

In water that is not deep relative to the amplitude of the tidal wave the duration of the rise is usually shorter than that of the fall. This effect has been explained by the consideration that the velocity of a wave varies with the depth with the result that the crest moves faster than the trough causing the wave front to steepen. Such an effect may be seen when an ocean swell approaches a beach and results finally in the breaking of the wave. It is illustrated in Figure VI-1 which shows a tide curve from Barnstable Harbor.

The distortion of the tide curve is attributed to harmonics having a period of some fraction of that of the M_2 constituent and designated as M_4, M_6, M_8, etc., the subscripts indicating the number of harmonic waves which occur per day.

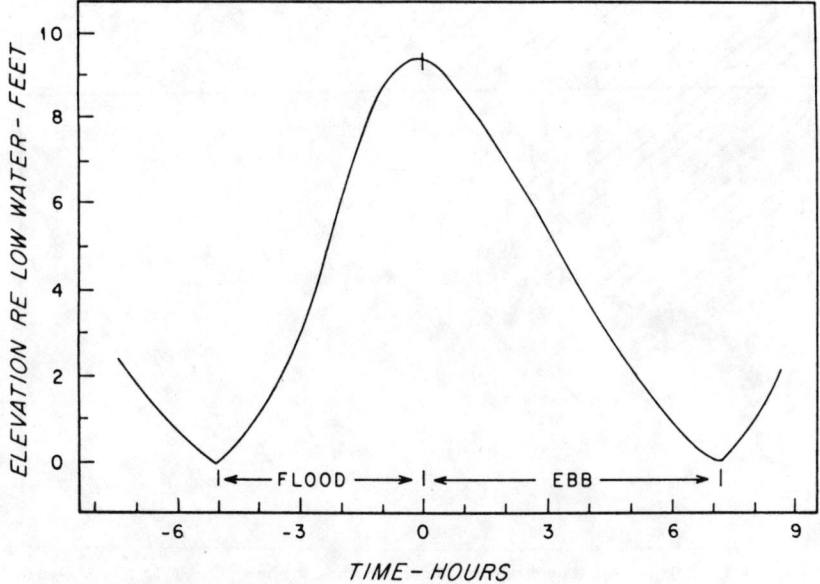

Fig. VI-1. Tide curve in Barnstable Harbor.

The predictions given in the tide tables do not indicate that the duration of the flood is always shorter than that of the ebb. For Long Island Sound, Vineyard Sound, south of Nantucket Island, in Cape Cod Bay, on Georges Bank, on parts of the Maine coast and in the Bay of Fundy the duration of the flood is shorter than the ebb of the tide. But in Narragansett Bay, Buzzards Bay, Nantucket Sound, on the outer coast of Long Island and much of the coast of the Gulf of Maine the duration of the flood is longer than that of the ebb (see Figure VI-2). This relation is clearly shown by the tide in Buzzards Bay, illustrated by a tide curve obtained at North Falmouth, shown in Figure VI-3, and by similar tide curves for Providence at the head of Narragansett Bay, shown in Figure VI-4. It cannot be explained by the relative velocities of the wave at high and low water as in the more usual case of shallow water tides.

A recent study of the harmonic analysis of the tide in Vineyard and Nantucket sounds has shown that the harmonics occur at nearly the same time and have nearly the same amplitude in all parts of this strait, while

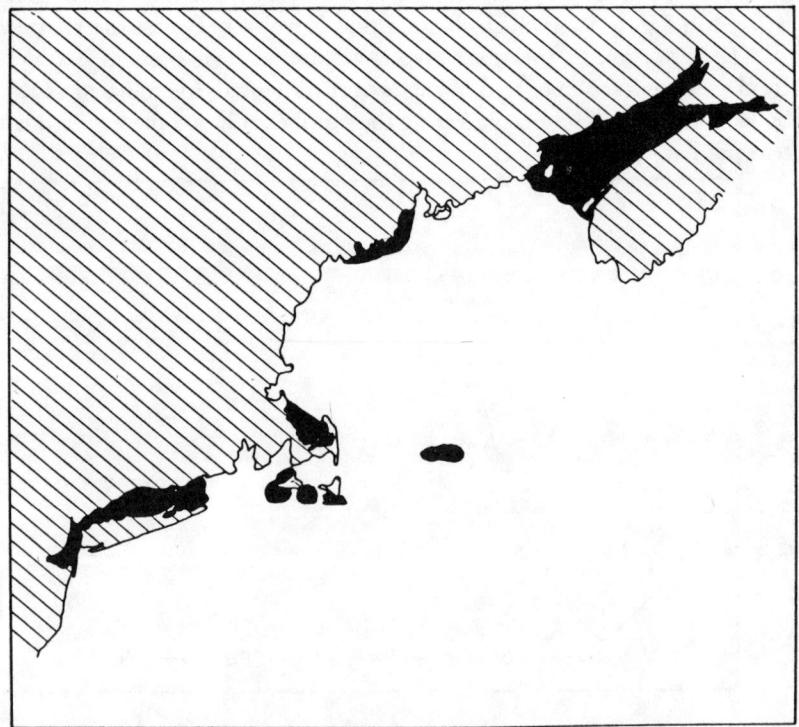

Fig. VI-2. Relative duration of rise and fall of tide on New England coast. In areas shown in black duration of rise is shorter than duration of fall. In other areas on coast duration of rise is longer than duration of fall.

Shallow Water Tides and Harmonics

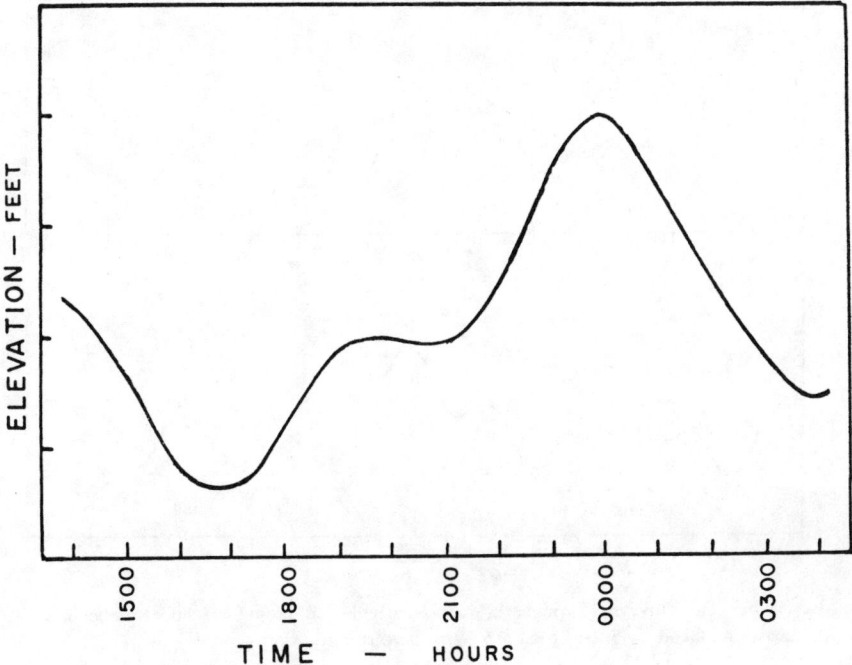

Fig. VI-3. Tide curve at North Falmouth. (Courtesy of John Moody.)

high water of the M_2 constituent, the principal component of the semidiurnal tide, occurs progressively later toward the eastern end of the strait. The result is that in Vineyard Sound the presence of the harmonics prolongs the period when the tide is falling but in Nantucket Sound this period is shortened. The change occurs at about Falmouth Heights near the junction of the sounds. Further study in other situations is needed to find the relation of the harmonics to the principal semidiurnal constituent of the tidal motion.

A correction for the effect of the presence of harmonics on the high water and slack water intervals of the predicted mean tide as given in the tide tables may be made from the following consideration. The asymmetry of the tide curve is attributed to the presence of harmonics. It may be removed by adding to the predicted high water interval, and subtracting from the predicted low water interval such a number as to cause the duration of the flood and of the ebb to be equal. The high water interval corrected for the presence of harmonics may be found from the following equation:

$$HWI\ (corrected) = \frac{HWI + LWI\ (as\ predicted) - 6.21}{2}$$

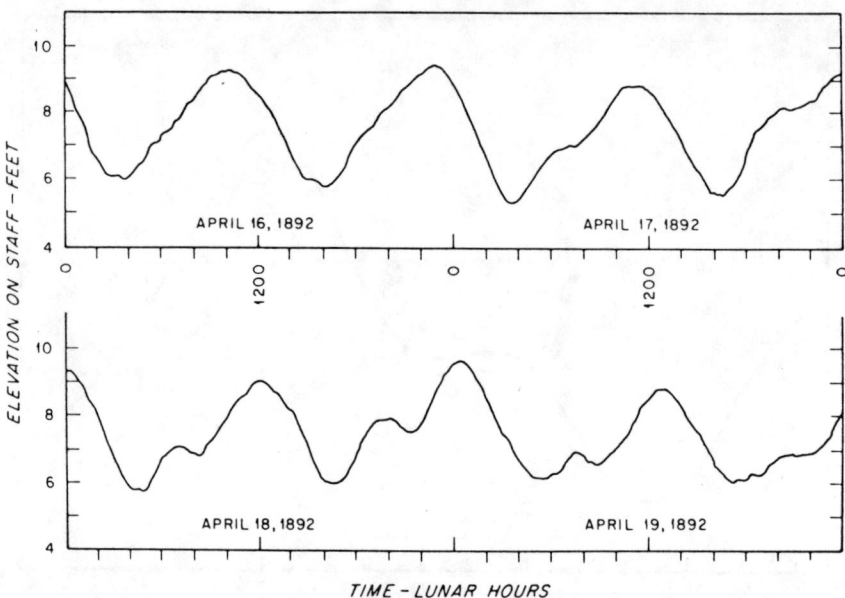

Fig. VI-4. Tide curves at Providence. (Modified after Harris. From Rept. Superintendent of U.S. Coast and Geodetic Survey for 1907. With permission.)

Similarily, in the case of the slack water interval,

$$SWI\ (corrected) = \frac{SWI\ before\ flood + SWI\ before\ ebb\ (as\ predicted) - 6.21}{2}$$

Wherever in the treatment of the tide in the following pages of this book the high water or slack water intervals are mentioned, they have been corrected for harmonics in this way.

Double high waters and *double low waters* are unusual conditions which occur respectively where the elevation of the tide shows two maxima separated by a period in which it is somewhat lower, or two minima separated by a period in which it is somewhat greater. The two conditions usually do not occur at the same place. The tide curve at Falmouth Harbor illustrates a typical double high water (Figure VI-5). It may be explained by the fact that the change in elevation due to the combined effect of the harmonics is maximal shortly before and shortly after the M_2 component is maximal so that their elevations are added to the elevation of the M_2 component. Midway between these two periods the elevation of the harmonics is minimal and results in an elevation when combined with that of the M_2 which is less than elevations reached either earlier or later.

Shallow Water Tides and Harmonics

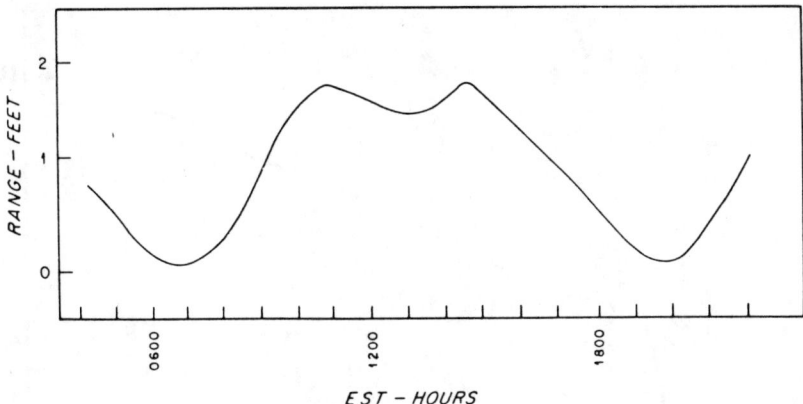

Fig. VI-5. Tide curve at Falmouth Harbor showing double high waters. (Courtesy of John Moody.)

TABLE 1
Greenwich Lunitidal Intervals at Reference Stations
(Supplied by the National Oceanographic and Atmospheric Administration)

Station	Mean High Water Interval, Hours	Mean Low Water Interval, Hours
St. John, N.B.	3.32	9.57
Eastport, ME	3.32	9.73
Portland, ME	3.60	9.72
Boston, MA	3.75	9.93
Newport, RI	0.25	5.85
New London, CT	2.08	8.52
Bridgeport, CT	3.77	10.13
Willets Point, NY	4.18	10.73
The Battery	0.98	7.32
Albany, NY	9.67	4.37
Sandy Hook, NJ	0.33	6.72

Mean Current Intervals - Hours

	Slack Flood Begins	Maximum Flood	Slack Ebb Begins	Maximum Ebb
Bay of Fundy Entrance	9.60	0.10	3.27	6.62
Portsmouth Harbor Entrance	11.93	1.77	5.30	8.08
Pollock Rip Channel	7.87	11.25	1.98	4.77
Cape Cod Canal - Railroad Bridge	8.25	11.07	2.22	4.67
The Race - Long Island Sound	10.82	1.07	4.28	7.08
Hell Gate - East River	9.03	11.87	2.63	5.58
The Narrows - New York Harbor	9.32	12.00	2.37	5.67

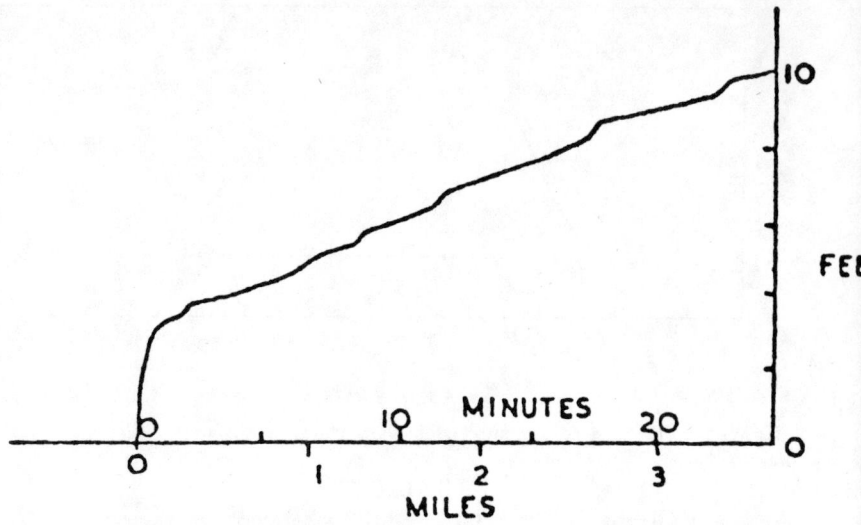

Fig. VI-6. Profile of bore in Petitcodiac River. (After Doodson and Warburg, Admiralty, Manual of Tides (1941). With permission.)

Fig. VI-7. Bore in Salmon River at Truro, Nova Scotia.

Shallow Water Tides and Harmonics

Double low waters have been reported to occur in Buzzards Bay at Marion, but this has not been confirmed by measurement. They may be present in the Cape Cod Canal at the railroad bridge where the predictions are discussed in the tide tables.

Tidal bores are extreme cases of the distortion of a tidal wave by shallow water. They occur in tidal rivers and bays which are shallow and in which the range in tide is exceptionally large. In such situations the water rises so rapidly that it cannot flow smoothly over the shallow bottom but reaches a velocity so great that its form becomes unstable and it advances as a breaking wave or series of breakers.

Tidal bores occur in the Petitcodiac River and other rivers which enter the Bay of Fundy near its head. The bore in the Petitcodiac River may be seen at Moncton, New Brunswick where its height is usually 2 or 3 feet, but during exceptionally high tides there may be a drop of 5 or 6 feet across the front. The tide does not enter the Petitcodiac River until about half tide, when it is rising rapidly, so that the bore passes Moncton about 3 hours before high water. It advances at a rate of about 10 miles per hour and the plunging water may be heard 10 minutes before its arrival. After the passage of the bore, the water continues to rise rapidly but smoothly (Figure VI-6). After high water, the water level falls rapidly for 3 or more hours, then decreases more slowly as the elevation is maintained by the river discharge.

A similar though less pronounced bore occurs in the Salmon River at Truro, Nova Scotia, where only 1 to 1.5 hours elapse between the passage of the bore and high water. It is a lively sight to see the bore pass. At low tide the bottom of the river is occupied by sand flats among which shallow channels wind carrying the river discharge to the sea. The bore moves steadily up stream now plunging across a sand flat, now shooting ahead more rapidly in the deeper channels where a series of small breaking waves form on the front to move slowly down its slope and to be replaced by others which form at a higher level (Figure VI-7).

The towns of Moncton and Truro provide good places to see the bores in the Petitcodiac and Salmon rivers and provide information when the next bore is expected. Those planning such a visit should select a day when the range of spring tides is large in order to see the bore at its best.

References

Doodson, A. T. and H. D. Warburg. Admiralty Manual of Tides. Her Majesty's Stationary Office, London, 277 pp., 1941.

Redfield, A. C. The tide in coastal waters. Journal of Marine Research. 36, 255-294, 1978.

Tricher, R. A. R. Bores, Breakers, Waves, and Wakes. American Elsevier Publication Co., New York. 250 pp., 1964.

Chapter VII

THE ROTATION OF THE EARTH

The rotation of the earth causes currents to turn to the right in the northern hemisphere as the result of what are known as *Coriolis forces*. In the southern hemisphere, which does not concern us, the deflection is to the left. The effect is illustrated by the currents on the continental shelf where the movement is not greatly influenced by land barriers. The direction of the current changes steadily, completing a change of 360° in each tidal cycle.

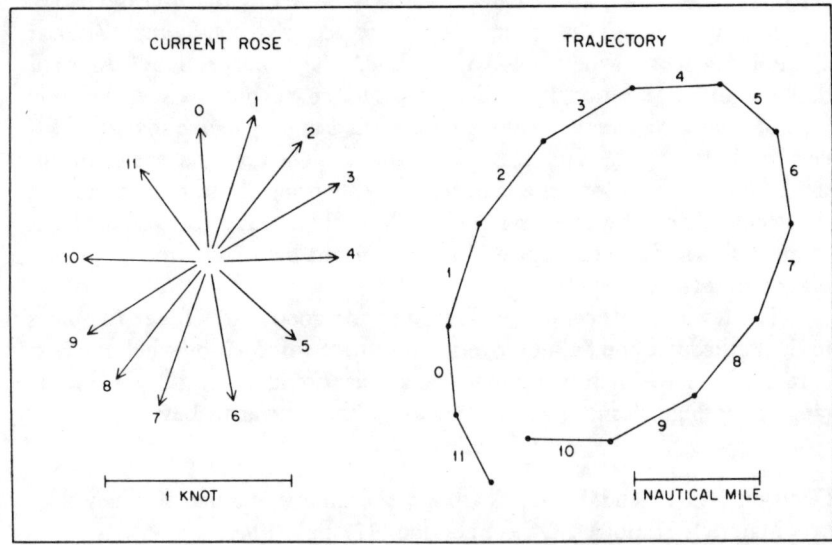

Fig. VII-1. Current at Nantucket Lightship. Left — current rose. Arrows show direction of current at periods of one lunar hour after transit of moon at Greenwich. Their length shows the velocity of the current at the time of measurement. Right — the trajectory of the current as deduced from the current rose.

The effect of the rotation of the earth is illustrated by the currents at the Nantucket Lightship which is anchored about 60 miles southeast of Nantucket Island. The current rose at the lightship is shown at the left in Figure VII-1. The direction of the current is indicated by arrows, the length of which is proportional to the velocity. The numbers indicate intervals of one lunar hour and show the progressive turning to the right during each interval. The drift of the water in the region of the lightship is given by its trajectory, obtained by plotting the arrows of the current rose end to end. The water may be seen to drift in a broad ellipse which does not quite close indicating the presence of a small nontidal current in a northeasterly direction.

Near the coast the water must flow nearly parallel to the shore and the elliptical form of the current rose is greatly altered. This is illustrated by the tidal currents at the Fire Island Lightship when it was anchored about 10 nautical miles south of the Long Island coast (Figure VII-2). The current rose

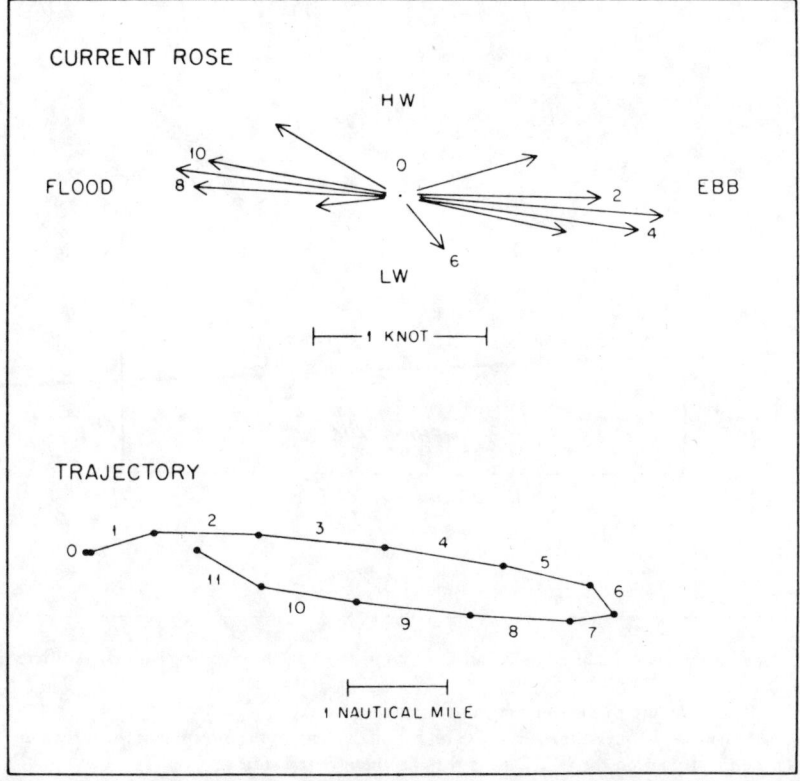

Fig. VII-2. Current at Fire Island Lightship. Above — current rose. Arrows show direction of current at periods of one lunar hour after transit of moon at Greenwich. Their length shows the velocity of the current at the time of measurement. Below — the trajectory of the current as deduced from the current rose.

is greatly flattened and the trajectory indicates that the water moves parallel to the coast except near the time when its direction is changing. The trajectory fails to close by about one nautical mile indicating a nontidal movement toward the east at this position.

Within embayments the shores prevent the current from turning under the influence of the rotation of the earth. Instead, water is forced toward the shore on the right side of the current until a gradient is produced in which the slope across the channel balances the effect of the Coriolis force (Figure VII-3A). As a result, the range is greater on the right of the flood current

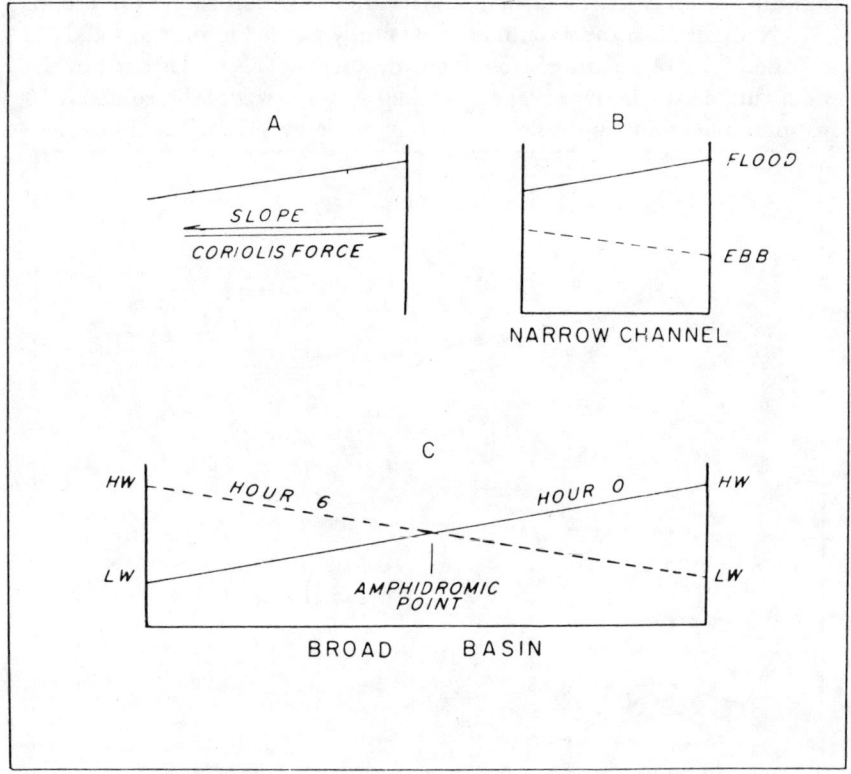

Fig. VII-3. Diagram of effect of earth's rotation on slope of surface in channels of different width.
A. The force due to the slope equals the Coriolis force.
B. In narrow channels the surface rises higher on the right of the direction of the current during the flood than on the left. During the ebb the reverse is true.
C. In broad basins, in which an amphidromic system is present, the surface slopes downward from one side where it is high water to the opposite side where it is low water. Near the center of the basin there is an amphidromic point where the elevation of the surface is at mean sea level and there is no tide.

and less on the opposite side (Figure VII-3B). The wave produced is known as a *Kelvin Wave*. This effect of the earth's rotation increases with the width of the passage and the speed of the flow.

In broad basins the rotation of the earth may give rise to an *amphidromic system* in which the time of high water increases along the margin of the basin. If the depth is appropriate, one period is required for high water to complete the circuit of the basin so that each high tide coincides with the next high tide to enter. At any time when it is high water on one side of the basin it is low water on the opposite side. Near the middle of the basin there is a position known as the *amphidromic point* where there is no tide (Figure VII-3C).

Figure VII-4 shows diagrammatically the effect of the width of the basin on the pattern of cotidal and corange lines developed under the influence of the earth's rotation. Cotidal lines indicate the positions where the times of high water coincide and corange lines show where the range is the same. If the channel is narrow, the earth's rotation has no effect on them and they are parallel (Figure VII-4A). In wider channels the earth's rotation causes the cotidal lines to advance relative to the corange lines on the right side of the flood current and to lag behind on the left side so that they cross at an angle (Figure VII-4B). When the basin is sufficiently broad to develop an amphidromic system, the cotidal and corange lines cross at practically a right angle (Figure VII-4C). Note that in an amphidromic system in the northern hemisphere the current turns toward the left rather than toward the right as in currents in the open sea.

In narrower basins, which are not sufficiently broad to permit the development of an amphidromic system, a so-called *degenerate amphidromic system* may develop in which the cotidal lines converge toward a virtual amphidromic point outside the basin (Figure VII-4D).

When the width of the channel is large, the theoretical equations based on interference phenomena cannot be used to indicate the behavior of the tide. The presence of such effects is shown by the large angle with which the cotidal and corange lines intersect, which may be as great at 90° in fully developed amphidromic systems.

On the coast of New England and New York most of the embayments and straits are too narrow to provide evidence from the predicted ranges and high water intervals for the effect of the rotation of the earth. In the Bay of Fundy, however, the cotidal and corange lines cross at an angle indicating that the width of the bay is sufficient for the rotation of the earth to affect the tide (Figure VII-5). Fully developed amphidromic systems are absent on this coast but one such system occurs in the nearby Gulf of St. Lawrence (Figure VII-6). An amphidromic system is also present in the deep water of the North Atlantic Ocean as illustrated in Figure X-1 (p. 63). A degenerate

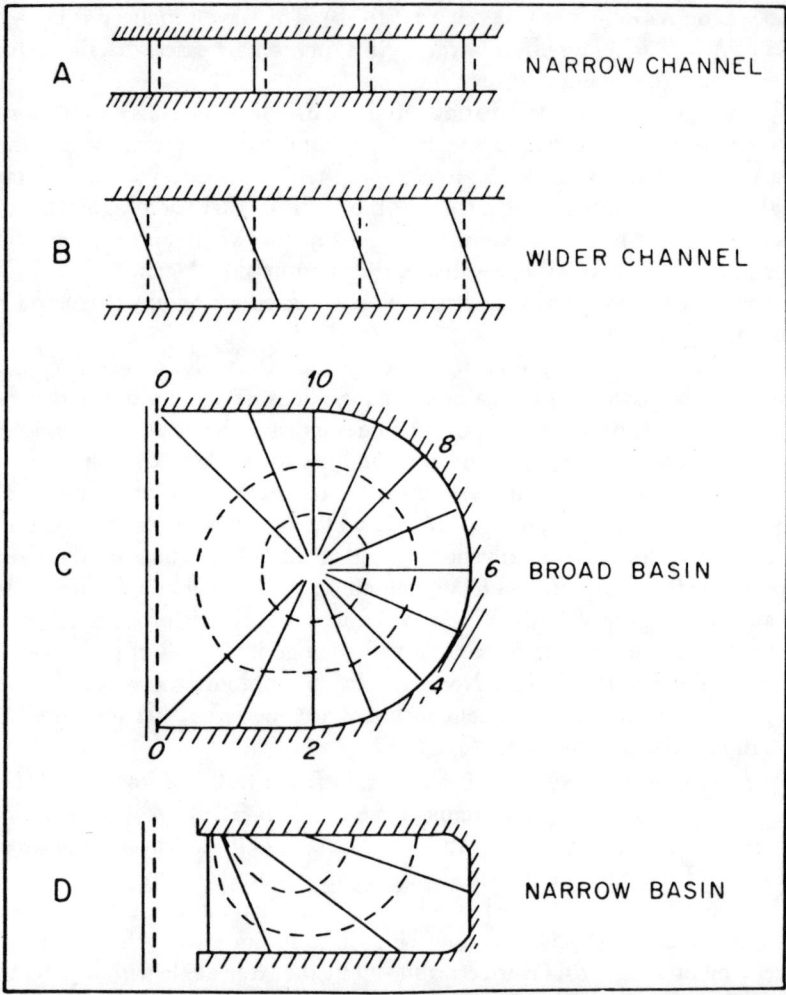

Fig. VII-4. Diagram of effect of earth's rotation on the relation of cotidal and corange lines in channels of different width. The solid lines are cotidal lines showing positions where high water occurs simultaneously. The dashed lines are corange lines showing positions where the range is the same.

A. In narrow channels the cotidal and corange lines are parallel, there being no significant effect of the rotation of the earth.

B. In wider channels the cotidal and corange lines cross at an angle indicating an effect of the rotation of the earth.

C. In broad basins, in which an amphidromic system occurs the cotidal and corange lines cross at an angle of about 90°.

D. In narrow basins, not sufficiently broad to develop an amphidromic system, a degenerate amphidromic system occurs in which the cotidal lines appear to converge at a virtual amphidromic point, located outside the basin, and cross the corange lines at a large angle.

The Rotation of the Earth

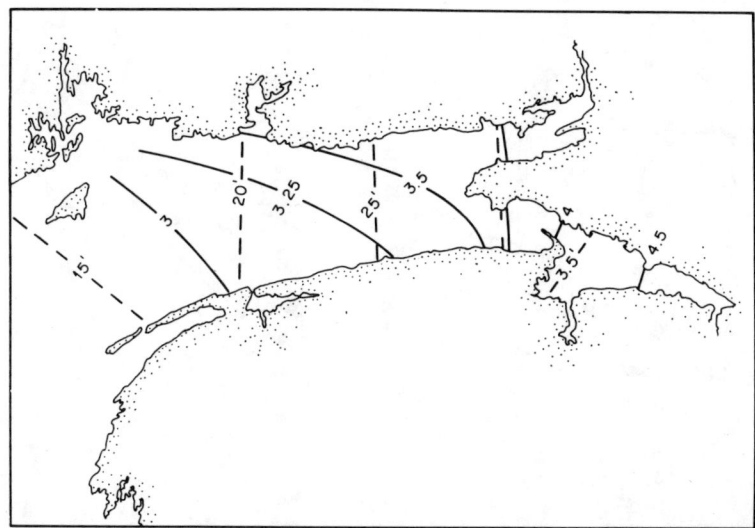

Fig. VII-5. The Bay of Fundy. Showing the angle at which the cotidal lines and corange lines intersect as evidence of the effect of the earth's rotation. Solid lines are cotidal lines, in hours, after moon's transit at Greenwich. Dashed lines are corange lines, in feet. (From the *Journal of Marine Research*, Vol. 36. With permission.)

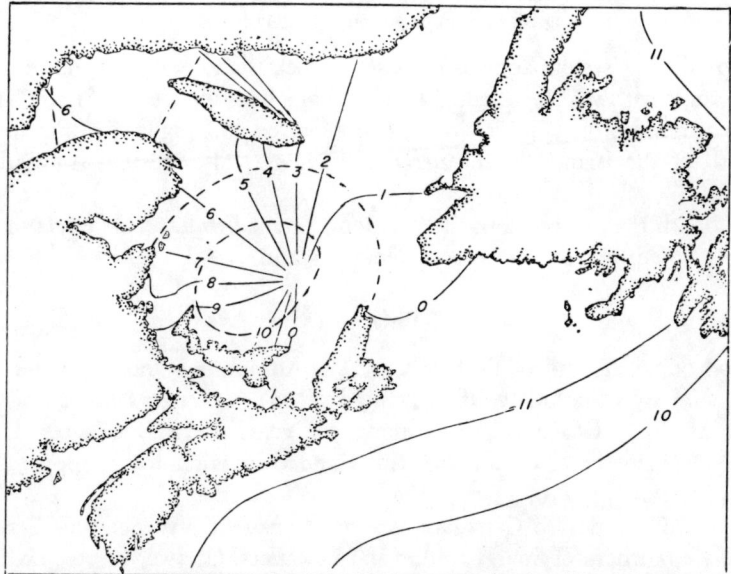

Fig. VII-6. Gulf of St. Lawrence. Showing amphidromic system with amphidromic point near Magdalen Islands. (In part from Rept. Superintendent of U.S. Coast and Geodetic Survey for 1904, and in part from Tides in Canadian Waters. With permission.)

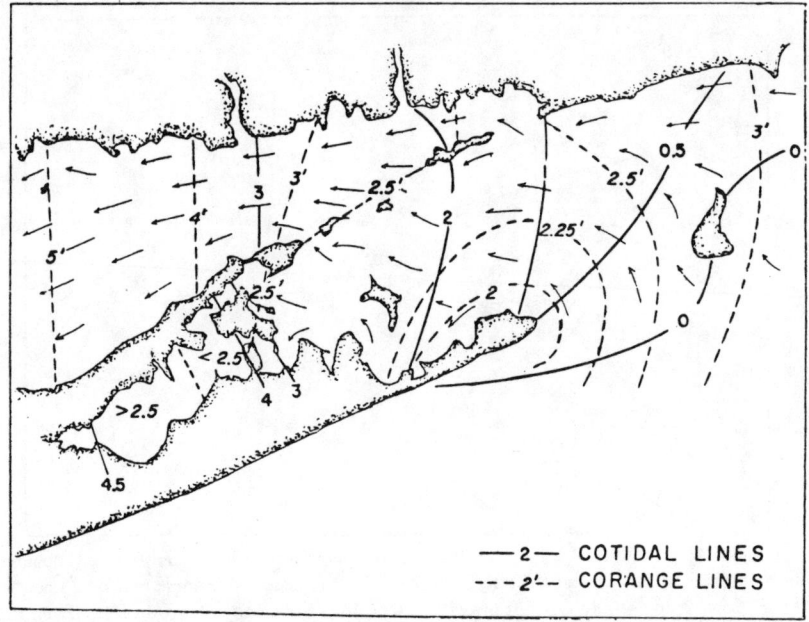

Fig. VII-7. Block Island Sound. Showing degenerate amphidromic system with virtual amphidromic point on Long Island or in ocean south of Long Island.

amphidromic system appears to exist in Block Island Sound as indicated by the angle with which the cotidal and corange lines intersect in Figure VII-7. The amphidromic point is virtual, being located on Long Island or in the sea south of the island. A degenerate system may also exist in the Gulf of Maine.

With these exceptions the rotation of the earth has little effect on the tide as predicted on the coast of New England.

References

Doodson, A. T. and H. D. Warburg. The Admiralty Manual of Tides. Her Majesty's Stationary Office, London, 1941. Chapters 20 to 25.

Haight, F. J. Coastal currents along the coast of United States. U. S. Department of Commerce, Coast and Geodetic Survey, Special Publication No. 230, 73 pp., 1942.

Dohler, G. Tides in Canadian Waters. Canadian Hydrographic Service, Department of Energy, Mines and Resources, Ottawa, 14 pp., no date.

Chapter VIII

METEOROLOGICAL EFFECTS ON THE TIDE

The predictions for the tide given in the tide tables for the reference stations are based on a long series of observations and are consequently quite precise. The difference between the tide at other positions and at nearby reference stations depends on measurements made over shorter periods, usually 29 days or its multiple, and are subject to small errors. Storms may cause departures from the predicted tide on the particular days on which departures occur which are referred to as *meteorological effects*.

The frequency and magnitude of departures of the tide from the predictions are shown in Table 2 based on observations at Portland, Maine, in May and November 1919, when there were 119 tides each month and where the mean range is 9.0 feet. It may be seen that the departures of the elevation of the tide at high water are larger in November than in May and are less frequent the greater they are. The differences in the time of high

TABLE 2
Errors in Prediction of Tides at Portland, Maine, for May and November, 1919. Mean range, 9.0 feet
(Tabulated after Marmer 1926, pages 206-208)

Error in Predicted Elevation of High Water	May	November
Maximum error, feet	0.9	1.9
Errors greater than 1.0 foot	0%	6%
Errors greater than 0.5 foot	11%	24%
Errors 0.5 foot or less	89%	70%
Error in Predicted Times of High Water	May	November
Maximum error, hours	0.4	0.4
Errors greater than 0.3 hour	5%	0.1%
Errors greater than 0.2 hour	35%	3%
Errors 0.2 hour or less	44%	97%
No error	15%	0%

water in the two months are less clear that that of the elevation. Although such effects are averaged out more or less completely in determining the mean tide, they may be significant on the days when they occur. Subsequent examinations have given results similar to these early studies, indicating that these give a reliable picture of the frequency and magnitude of meteorological effects on the tide.

Setup

The change in elevation due to meteorological effects, the *setup*, is measured relative to the predicted elevation at the time. It has two origins: 1) that due to the barometric pressure, and 2) that due to the wind. Both may occur at the same time.

The barometric effect is due to the fact that the sea acts as an inverted barometer. When the atmospheric pressure falls over a local area, the water rises until the weight of the added water balances the decreased weight of the air over the region, and vice versa if the atmospheric pressure rises above

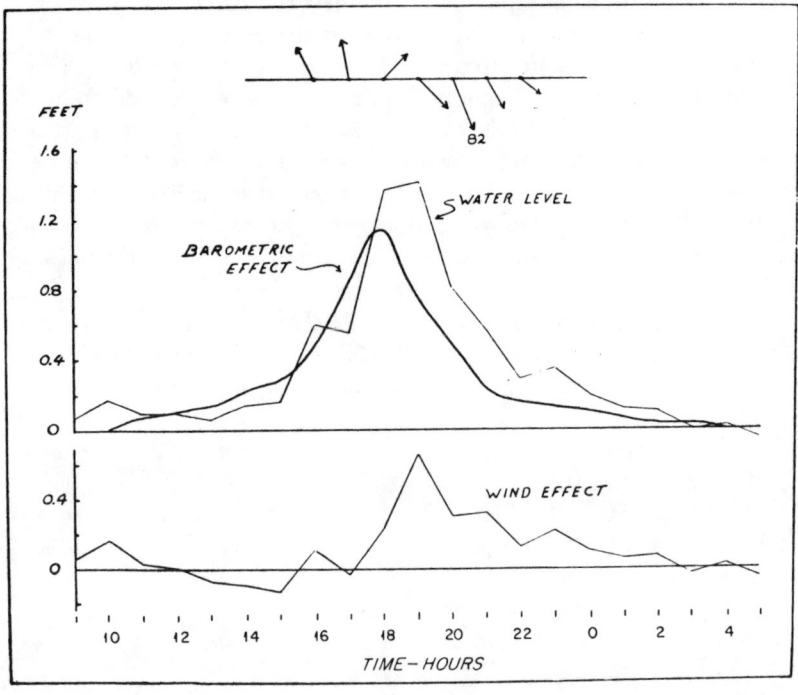

Fig. VIII-1. Bermuda. Hurricane of October 7, 1948. Ordinate: Water level at Ferry Reach and equivalent barometric effect estimated from barometer at Hamilton. Abscissa: Atlantic standard time. Above: Coincident wind direction and speed (miles per hour). (From Meteorological Monographs, Vol. 2, No. 10. With permission.)

Meteorological Effects on the Tide

that of the surrounding region. Because the density of mercury is about 13.1 times as great as that of sea water, a fall in the barometer of 1 mm Hg will cause the sea surface to rise about 0.043 feet, or a fall of 1 inch in the barometer will cause a rise in sea level of about 1.1 feet.

Figure VIII-1 shows simultaneous measurements of the water level and the estimated barometric effect during the passage of a small hurricane at Bermuda. During the early stages of the storm when the wind was from a southerly direction, passing over deep water in approaching the island, the rise in water level was about that expected from the barometric effect. There was no clear evidence that the wind had any effect on the rise in the water. After the passage of the storm center, when the barometer commenced to rise and the wind shifted to a northerly direction, the water level continued to rise and remained above that attributed to the barometric effect until the storm had passed. During this period when the wind blew across the shallow water of the lagoon, it had a significant effect upon the level of the water.

The *wind setup* is the increase in water level due to the wind. It arises because the wind exerts a drag on the water causing the water to move in the

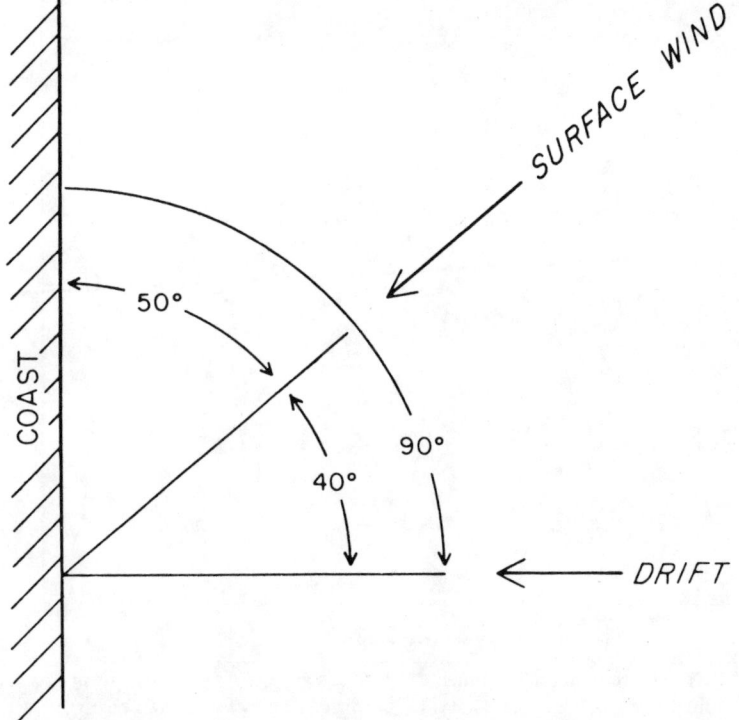

Fig. VIII-2. Relative direction of coast, surface wind and surface drift of water to produce maximum wind set up.

general direction of the wind. On reaching a coast, the motion due to onshore winds is checked and water accumulates there until an equilibrium is reached between the force which produces the accumulation and the dispersive force arising from the resulting slope of the water level. The wind setup may be measured by the water level at which this equilibrium occurs after correcting for the barometric effect.

When the wind is offshore, the setup is negative, the water level being less than predicted. Such negative setup is sometimes known as *set down*.

The setup which accompanies strong winds on the outer coast of the western North Atlantic Ocean has been studied at Atlantic City, New Jersey, by comparing the elevation of the water recorded in excess of that predicted for the time, after correcting for the barometric effect. The surface wind was

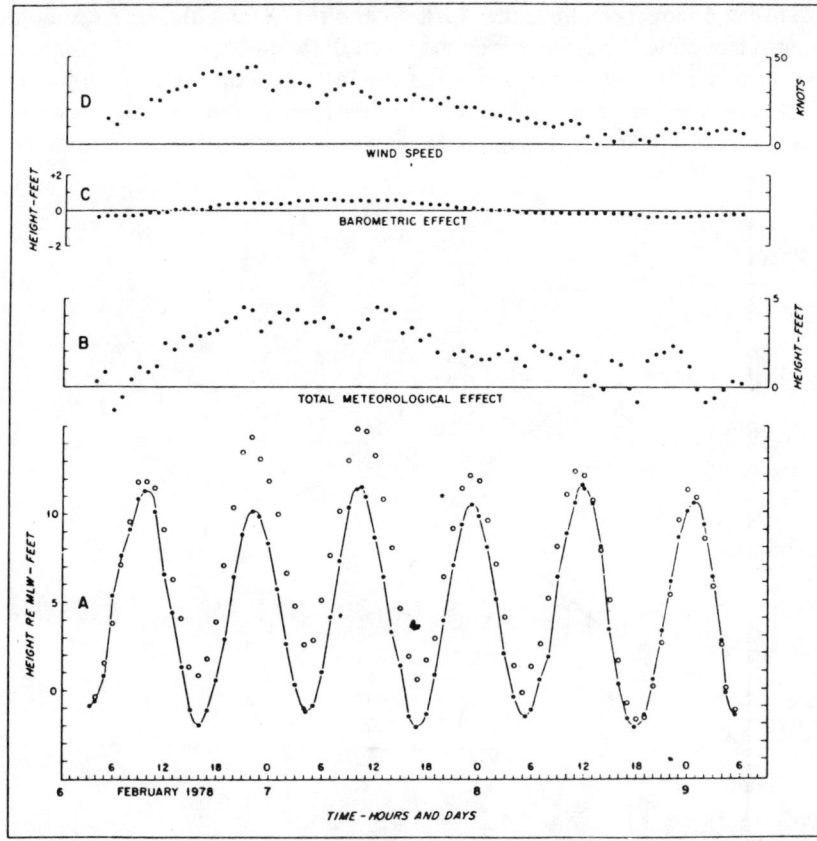

Fig. VIII-3. Boston. Storm of February 1978. Abscissa: Eastern Standard Time in days and hours. Ordinates: A, curve through solid circles predicted height of sea surface relative to mean low water. Open circles, observed height at Commonwealth Pier. B, difference between height observed and height predicted. C, estimated effect of barometric pressure on height of sea surface. D, wind speed at Boston International Airport.

Meteorological Effects on the Tide

deduced from the barometric pressures shown on the weather maps. It was concluded that the drift of the water was about 40° to the right of the direction of the surface wind. Since the maximum setup may be expected to occur when the drift is at 90° to the coast, it will occur when the surface wind approaches the coast at an angle of about 50° (Figure VIII-2). The maximum setup at Atlantic City was 0.044 to 0.050 foot per knot of surface wind speed. These are the relations to be expected when the wind blows steadily in the same direction for at least 10 hours. They are subject to some error because of uncertainties in the theoretical relations between the barometric gradients, the surface wind and the drift of the water.

Extra high sea levels occur every winter along the coast as the result of severe storms which develop along cold fronts. Occasionally they have an elevation which causes serious damage. The most famous of these storms was the Portland Gale of November 1898, notable because of the loss of life due to the sinking of the steamship *Portland*. The surge accompanying this storm breached the beach at the mouth of North River and damaged many structures along the shore. Such a storm in February 1978 caused local damage on the shores of Massachusetts Bay equivalent to that produced by a tropical hurricane.

The tide gauge at Boston (Commonwealth Pier), constructed so as to eliminate the short period fluctuations in sea level such as are produced by the wind waves and to show only the longer period tidal changes, showed that prior to noon on February 6, 1978, the sea level fluctuated about that predicted. It then rose rapidly until about 10:00 p.m. when the highest level of 14.39 feet above mean low water was reached. This was 4.5 feet above that predicted. The difference between the recorded and the predicted sea level remained large as the tide fell and rose, fluctuating between 2.8 and 4.5 feet until noon on February 7, a period of 14 hours. It then fell slowly for the remainder of February 7 (see A and B, Figure VIII-3). The rise in sea level due to this storm lasted very much longer than the surges which usually accompany tropical hurricanes and ordinarily last for only 4 or 5 hours.

The extra elevation of sea level was due to both the fall in barometric pressure and to the effect of the wind. The barometric effect was relatively small (Figure VIII-3). The lowest barometric reading was at 6:00 a.m. on February 7 and corresponded to an extra elevation of 0.65 foot or about 15 percent of the maximum extra elevation occuring during the storm. The course of the extra elevation of the sea followed closely that of the recorded wind speed (compare B with D, Figure VIII-3). The setup due to the wind during this storm was about 0.085 foot per knot which is somewhat greater than that calculated for Atlantic City. During the storm the wind was gusty, the gusts having a speed up to 50 percent greater than the recorded steady speed, a fact which may make such calculation difficult.

Before and during the maximum winds their direction was from a little to the east of north. Presumably the direction of the drift of the water produced was from the northeast. As the wind diminished, it veered about 30° to the right. The major physical damage due to this storm was probaby caused by the wind waves which developed on the raised sea surface. This was greatest on beaches exposed to the fetch of the wind across Massachusetts Bay and occurred at elevations above that of the rise in sea level due to the wind, as at Nantucket Beach and Scituate where buildings had been built too close to the level of the sea.

In enclosed bodies of water, in straits and embayments, the trend of the coast may differ greatly on opposite sides of the enclosure and from that of the outer coast. The effect of the wind on the setup in such a region has been studied in Nantucket Sound by examining the tide gauge records at Nantucket, Harwichport, West Chop, and Woods Hole (Figure VIII-4). The results showed that the departures from mean sea level could be resolved into two components: that which was a response to the wind direction on the outer coast and that due to the local wind at each tide station. At different positions these components vary in relative importance.

When the setup determined for the four stations considered was

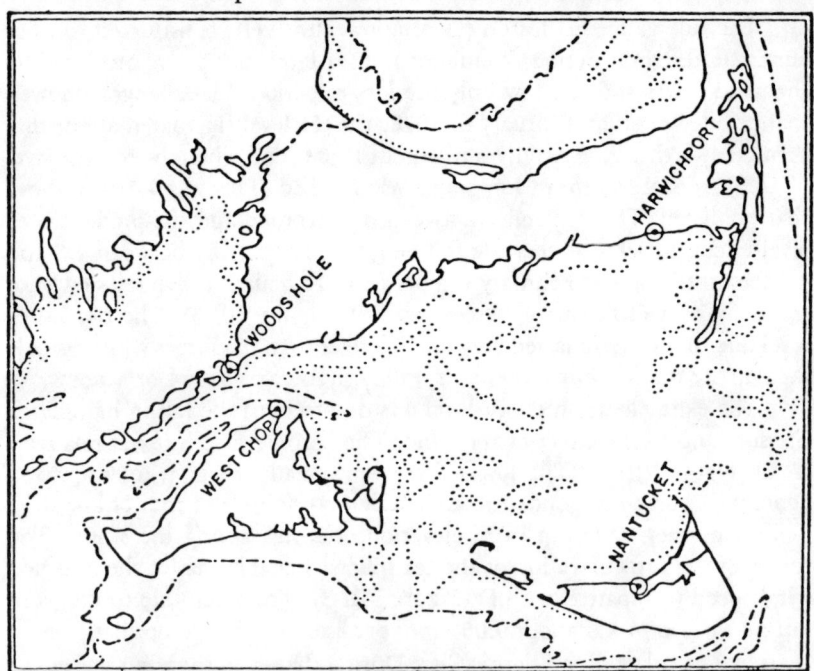

Fig. VIII-4. Nantucket and Vineyard Sound, showing positions of tide gauges at Harwichport, Nantucket, West Chop, and Woods Hole. (From *Limnology and Oceanography*, Vol. 3. With permission.)

averaged, the results showed that maximum setups occurred with northeasterly winds and negative setups with southwesterly winds. They may be attributed to the effect of the wind on the water of the outer ocean which influences the sound as a whole. At each station the local setup is also influenced by the direction at which it is approached by the wind. Nantucket is protected by land on the east and south sides and the local setup is limited to the fetch of the wind across Nantucket Sound. Maximum local setups occur with northwesterly winds and negative setups with easterly winds. The net effect of the combination of these two components is a maximum setup with northerly winds and negative setups which are greatest with southerly winds. Harwichport is situated on the side of Nantucket Sound opposite to Nantucket. It is protected by land from northerly winds. The maximum local setup occurs with southwesterly winds which have a fetch across Nantucket Sound and negative local setups occur during winds from the north-northwest. The net effect of the combination of the components due to the wind on the water of the outer coast and the local setup is to produce a maximum setup with northeasterly winds and a negative setup with winds from the west. The trend of the coast at Harwichport does not differ greatly from that of the outer coast which is perhaps why the setup there is much like that attributed to the outer ocean. The water levels at West Chop and Woods Hole are influenced by the flow through Vineyard Sound, and possibly in the case of Woods Hole from Buzzards Bay, and need not be discussed.

The local setup depends so much on the local topography, the speed of the storm and the local fetch that useful generalizations cannot be stated regarding the setup to be observed. Each position must be examined with respect to the local conditions.

Hurricanes

Hurricanes are cyclonic storms which develop in the tropical Atlantic. The wind blows with great speed turning in an anti-clockwise direction about a localized depression in atmospheric pressure, or center, as the storm moves across the surface of the ocean. Hurricanes cause a rapid rise in water level, or storm surge, typically with a duration of not more than several hours in contrast to a setup which is a relatively slow rise or fall with a duration of many hours or days depending on the persistence and direction of the wind. The diameter of the disturbance may be 300 or 400 miles and wind velocities near the center exceed 75 miles per hour and in the gusts may be greater than 100 miles per hour.

Hurricanes originating in the tropical Atlantic Ocean move in a westerly direction until the West Indies are approached when they may continue westward or turn northward to cross the coast south of Cape Hatteras. Some hurricanes, veering toward the east, clear the coast and proceed into the North Atlantic. Along the coast these give rise to northerly autumnal gales, sometimes known as equinoctial gales. Only a few cross the New England coast, but when they do they are of full hurricane intensity.

The first recorded hurricane to cross the New England coast occured in August 1635. According to William Bradford's diary, the water rose 20 feet at the head of Buzzards Bay. Severe hurricanes crossed the coast in 1805 and 1821. Study of the latter established the cyclonic character of hurricanes. A storm in 1879 which caused great damage to ships at Providence was probably such a hurricane. Recent hurricanes which have crossed the New England coast did so on September 21, 1938, September 14, 1944, August 31, 1954 (Carol), and September 11, 1954 (Edna). The paths of the center of these hurricanes are shown in Figure VIII-5.

The changes in water level at the coast may be discussed under three categories, the hurricane surge, forerunners, and resurgences.

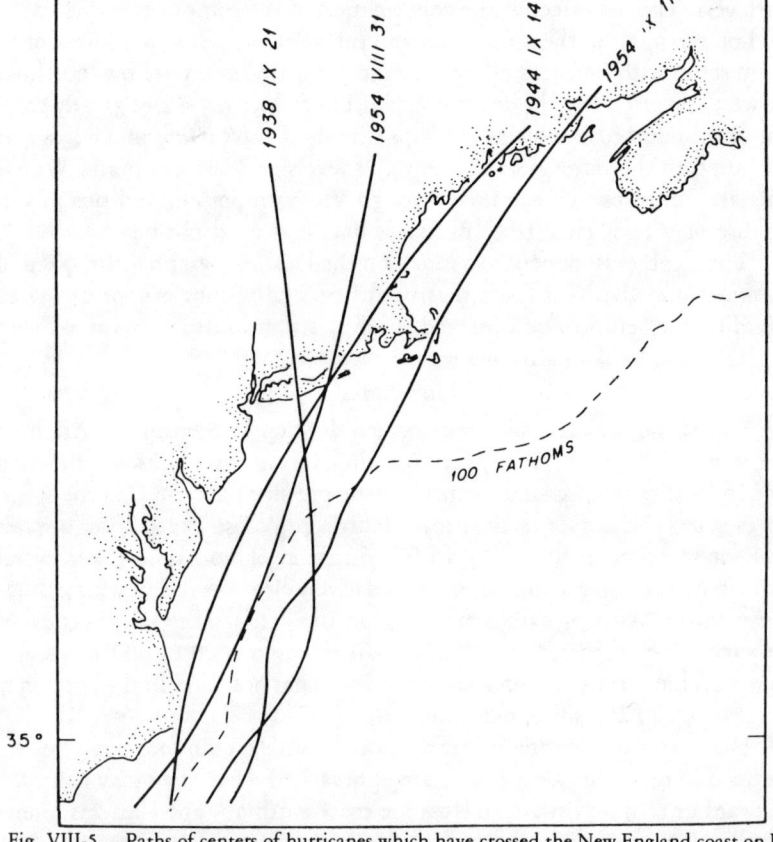

Fig. VIII-5. Paths of centers of hurricanes which have crossed the New England coast on IX 21 1938, IX 14 1944, VIII 31 1954, and IX 11 1954. (From *Meteorological Monographs*, Vol. 2, No. 10. With permission.)

Meteorological Effects on the Tide 53

The *hurricane surge* is a very rapid rise in water level and an almost equally abrupt fall which occurs at the time of the passage of the storm across the coast. The water level may rise as rapidly as 2 or 3 feet per hour and to elevations 10 or 15 feet above the predicted tide. The floods produced do not usually last for more than 4 or 5 hours. The relation of the maximum water level to the distance from the center, or eye, of the storm is shown in Figure VIII-6 which is based on tide gauge records and reliable field observations along the outer coast. The maximum levels do not occur at the center of the storm but 40 to 100 miles east of it. On either side of the maximum the levels decrease, becoming insignificant 200 miles to the east of the center and 100 miles to the west of it.

These relations are to be understood by a consideration of the wind field about the center of a hurricane as it approaches and crosses a coast. In their cyclonic motion about the storm center the winds on the right of the center tend to blow in the direction of its motion and move the surface water in that direction. In the deep water the waves produced by the wind move more rapidly than the storm so that little or no change in the water level is produced by the wind, as was the case of a hurricane at Bermuda. On crossing the shallower water of the continental shelf, the wave of wind-driven water is reduced in velocity and does not move faster than the storm. The wind-driven water consequently accumulates in the storm's path to form a solitary wave which on reaching the coast produces a surge of the height shown in Figure VIII-6. On the left of the storm center, on the other hand,

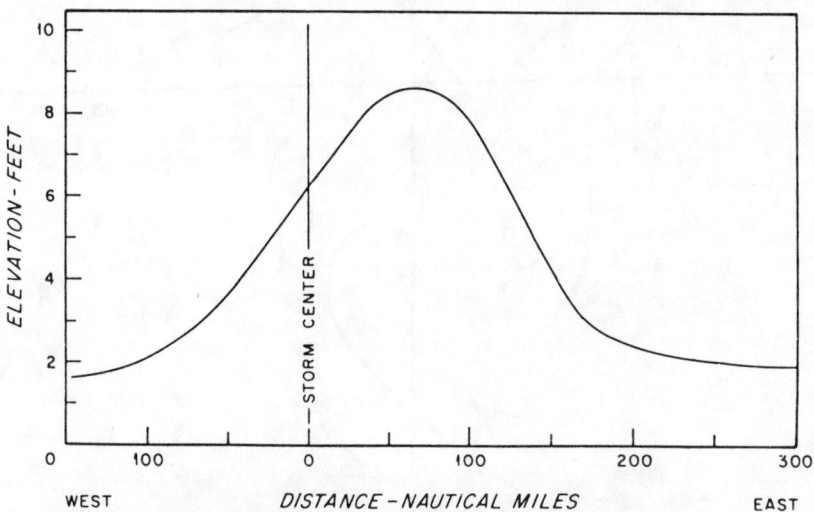

Fig. VIII-6. Maximum water levels during hurricanes above predicted tide along outer coast of southern New England relative to distance from storm center. Ordinate: elevation in feet. Abscissa: Distance in nautical miles. (Adapted from *Meteorological Monographs*, Vol. 2, No. 10. With permission.)

the winds tend to blow in a direction opposite to that in which the storm is moving. Little or no accumulation of water occurs on this side of the storm's

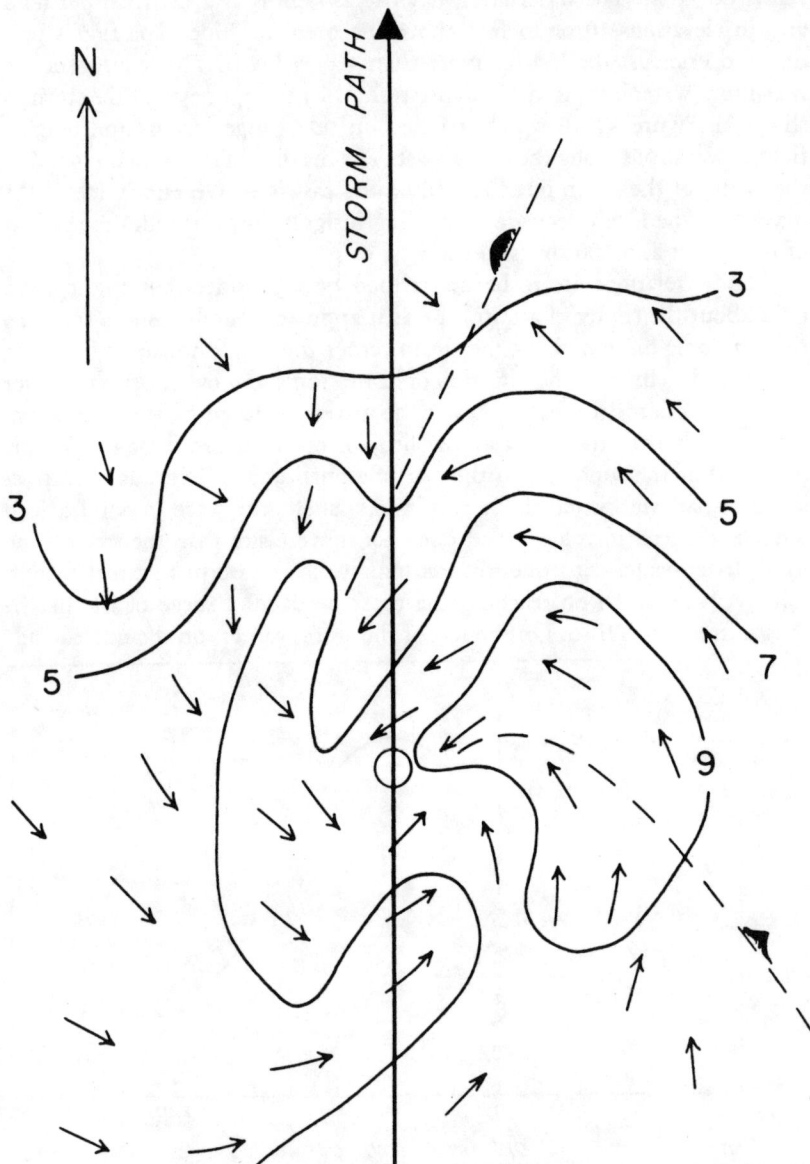

Fig. VIII-7. Wind field about center of hurricane of September 26, 1936 as it passed New Haven, Conn. Arrows show direction of wind. Contours its velocity on the Beaufort scale. (From *Meteorological Monographs*, Vol. 2, No. 10. With permission.)

path. Figure VIII-7 shows the wind field of the hurricane of September 1938 as it crossed the coast at New Haven, Connecticut.

On entering closed embayments such as Narragansett Bay and Buzzards Bay, the hurricane surge is augmented by reflection from the head of the bay much as is the ordinary tide. The water level rose to about 15 feet above the predicted tide at the head of these bays during the hurricane of August

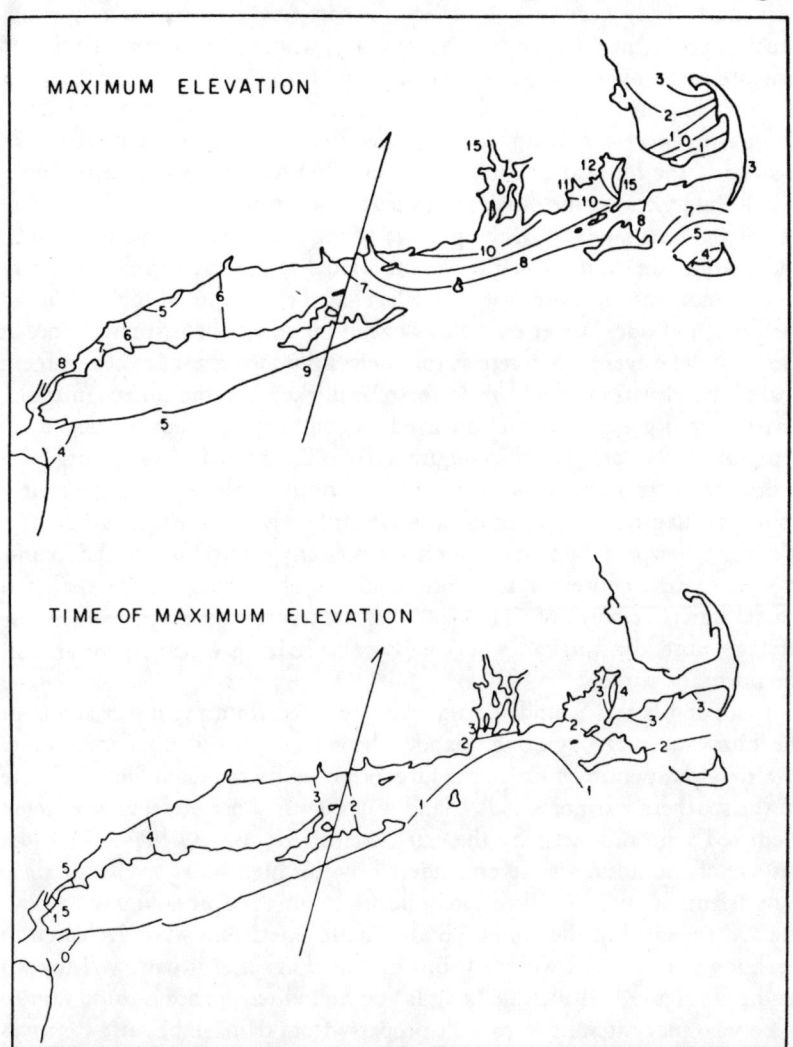

Fig. VIII-8. Hurricane of August 31, 1954. Above: Maximum elevation of water above that predicted in feet. Below: Time of maximum elevation, Eastern Standard Time in hours. Arrows show path of storm center. (Adapted from *Meteorological Monographs*, Vol. 2, No. 10. With permission.)

31, 1954. Figure VIII-8 shows the maximum water levels above their predicted values and the time of their occurrence.

Forerunners are small rises in tidal level above that predicted which develop many hours before the arrival of the hurricane surge. They are not greater than one or two feet. They have been attributed to the transport of water by the longer swells which have been known to precede approaching hurricanes. They may be due at least in part to the strong easterly winds which precede the advance of the storm and which cause a setup similar to that produced by winter gales. Forerunners vary greatly in different storms and may appear at positions where the hurricane surge is not recorded.

Resurgences are disturbances in water level which occur after the passage of the hurricane center when the wind has changed its direction.

Resurgences on the outer coast occur following the surge which accompanies the hurricane. With the passage of the storm center, the wind shifts and permits the water which has accumulated at the coast to flow seaward and its momentum causes the water level at the coast to fall below that of the predicted tide. Water then flows back to produce the first resurgence at the coast. The events then repeat themselves producing a series of resurgent peaks, the elevation of which decrease until they become unrecognizable. Coastal resurgences which occurred accompanying the hurricane of September 1944 are shown in Figure VIII-9. The period of the resurgences appear to be related to the width of the continental shelf being about 5 hours at Atlantic City and nearly 8 hours at Sandy Hook where the shelf is wider. At Newport the first resurgence was clearly shown but not the following ones. Resurgences of this sort were not observed in more sheltered coastal waters or north of Cape Cod. Coastal resurgences do not add to the damage caused by hurricanes because they cause levels which are lower than the hurricane surge.

In Long Island Sound the water is protected from the direct action of the hurricane surge by Long Island. The maximum disturbance of water levels is almost entirely due to a resurgence from Block Island Sound. There on the southern coast of Rhode Island water levels about 10 feet above those predicted were produced by the hurricane of August 31, 1954. With the passage of the storm a wave engendered by this high water level moved up Long Island Sound, in a direction opposite to the then prevailing wind, and reached the head of the Sound about 3 hours later. This wave decreased in elevation as the sound widened, but increased again as it narrowed toward the head (Figure VIII-8). The Long Island Sound resurgence is important to those who may attempt to protect property from damage because it arrives long after the passage of the storm center when the occurrence of greater flooding may not be expected.

In Buzzards Bay the strong southeasterly winds which prevail during

Meteorological Effects on the Tide

the advance of the hurricanes raise the water level along the northwestern side of the bay above that predicted for the tide. As the center of the storm passes, the wind veers to the west. Water piled up on the northwest side of the bay is released and driven by the strong westerly wind surges across the bay to increase the level on the southeast side. For example, during the hurricane of August 31, 1954, the water level rose about 12 feet above that predicted along the northwestern side of Buzzards Bay. More than an hour later it rose to nearly 15 feet on the southeast side (Figure VIII-8). A similar resurgence was observed at Woods Hole during the storm of September 1938 when water flooded from Buzzards Bay to Great Harbor causing loss of life and major physical damage.

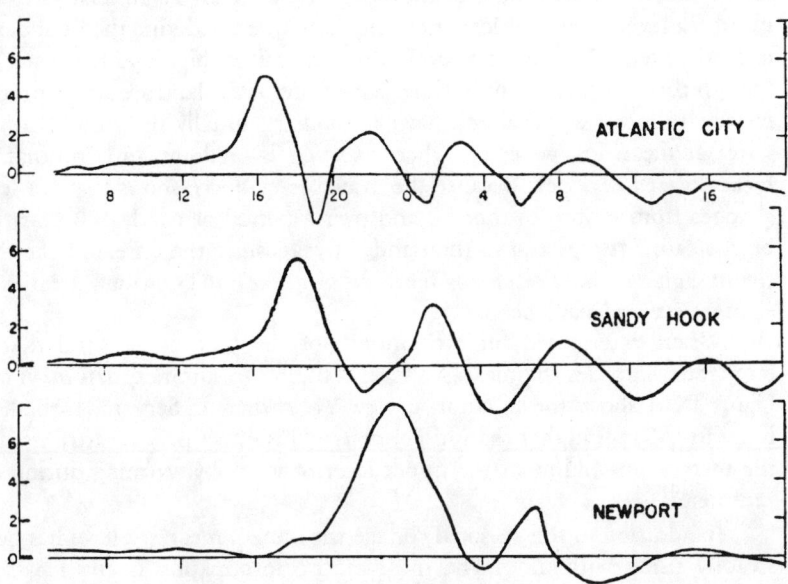

Fig. VIII-9. Coastal resurgences. Hurricane of September 14-15, 1944. Ordinate: Elevation above that predicted in feet. Abscissa: Eastern Standard Time, hours. (From *Meteorological Monographs*, Vol. 2, No. 10. With permission.)

References

Marmer, H. A. The Tide. Appleton, London and New York, 282 pp., 1926.

Miller, A. R. The effect of steady winds on sea level at Atlantic City. Meteorological Monographs, 2(10) 24-32, 1957.

Miller, A. R. The effect of winds on water levels on the New England coast. Limnology and Oceanography, 3(1) 1-14, 1958.

Cline, I. M. Tropical Cyclones. Macmillan, New York, 301 pp., 1926.

Redfield, A. C. and A. R. Miller. Water levels accompanying Atlantic Coast hurricanes. Meteorological Monographs, 2(10), 1-23, 1957.

Chapter IX

SEA LEVEL

In the theoretical treatment of the tide, elevations are given relative to the mean sea level. Laws and legal documents frequently define the boundaries of land in terms of mean sea level or its derivatives, high and low water, as though these were constant in time. Elevations of the land are also expressed relative to mean sea level and navigation charts usually show the depth of water at mean low water or, where the tide is predominantly diurnal, at mean lower low water. Because the available evidence shows that sea level changes from month to month, and over a period of nearly 100 years has been steadily rising, and in thousands of years since the retreat of the continental glaciers has risen many feet, it is of interest to know how great these changes are and have been.

When determined for each month of the year, it is found that in temperate latitudes the mean sea level is higher in summer than in winter. Figure IX-1 shows the situation at New York where in September the level is about 0.6 foot higher than in February. This effect may be attributed to the increase in volume of the upper layer of water by warming during the summer.

In addition to this seasonal change the annual mean sea level has been steadily rising with time. The most precise information is given by tide gauge records which have been kept at New York since 1893 and at Boston since 1922. These are shown in Figure IX-2. Very similar observations have been made at other positions along the Atlantic coast of the United States. Sea level relative to the land has been rising steadily on this coast at an average rate of about 0.01 foot per year during the past 50 years. Presumably, it may be expected to continue to rise, but whether it will do so remains to be seen.

Over longer periods of time, measured in thousands of years since the retreat of the continental glaciers, the rise in sea level has been much greater. These estimates, however, are subject to a larger error being based on the age determined by the radiocarbon content of organic relics believed to have been formed or deposited above or at the high tide level.

Sea Level

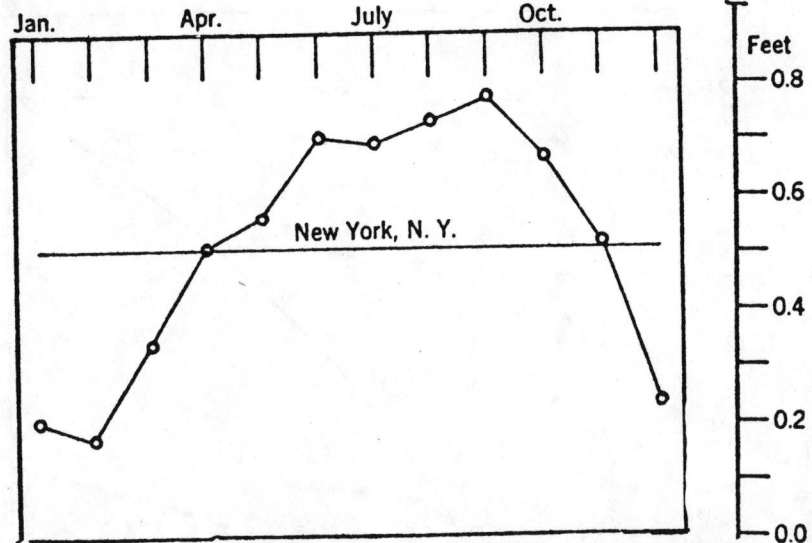

Fig. IX-1. Seasonal change in mean sea level. Reproduced from Meany, 1949.

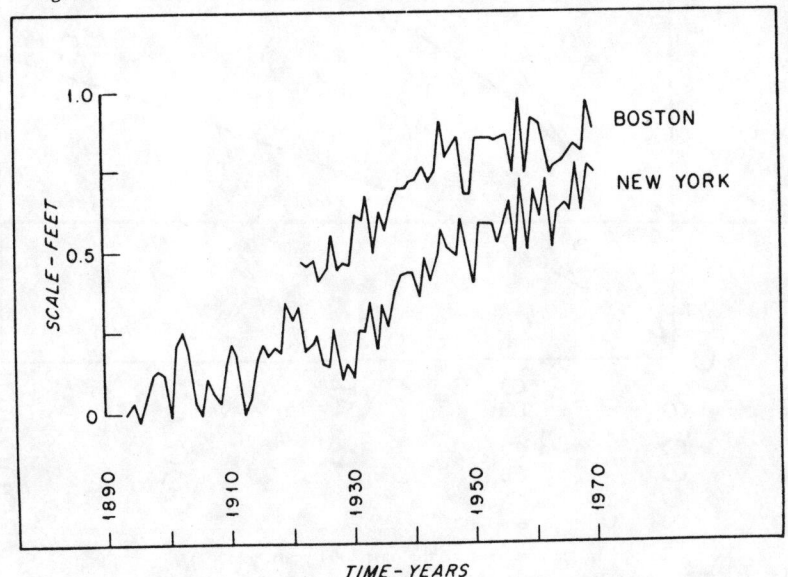

Fig. IX-2. Yearly variation in mean sea level at Boston and New York. Modified after Swanson, 1974.

The occurrence of fossil bones and teeth of land animals on the continental shelf is convincing evidence that the sea level has risen since glacial times. The recovery of the teeth of mammoths and mastodons by fishermen

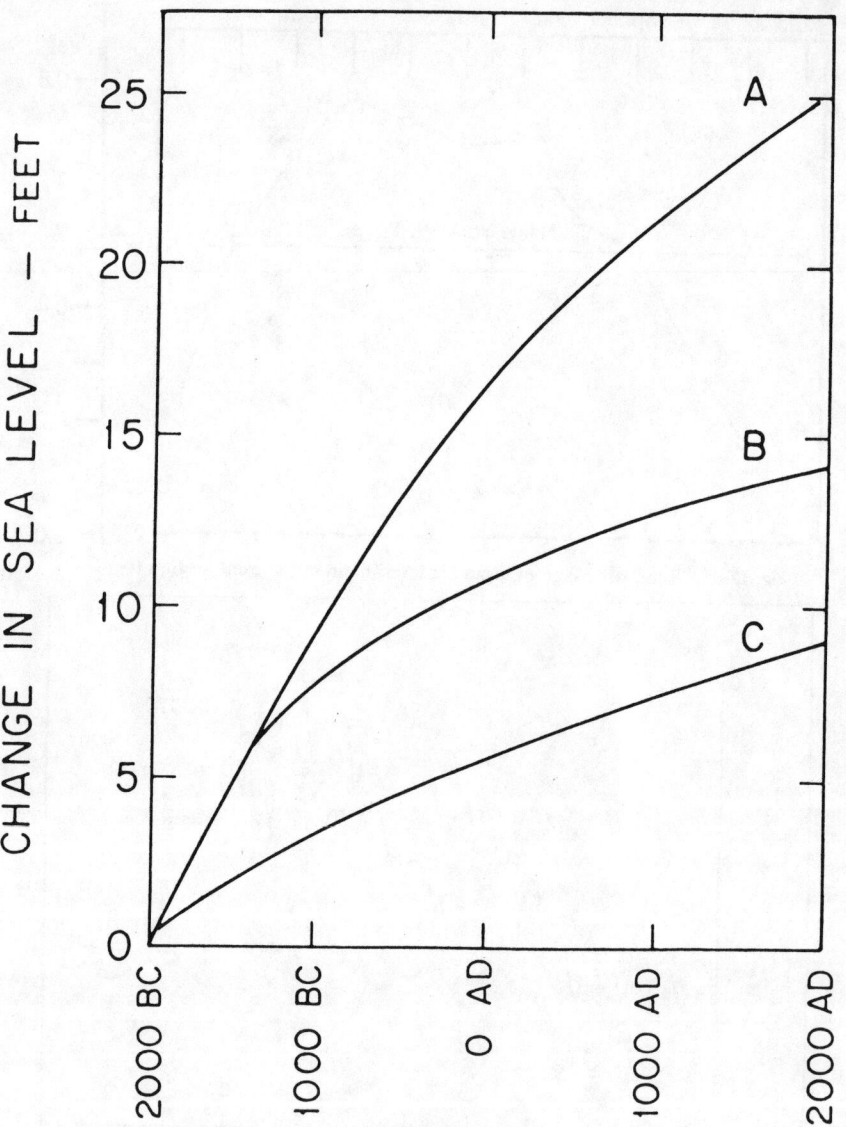

Fig. IX-3. Change in sea level during the past 4000 years.
A. On Cape Cod, Nantucket, and Long Island.
B. North of A at Newport and Plum Island.
C. South of A at Bermuda, North Carolina, Florida, and Louisiana. (In part from Redfield, 1967.)

from Geoges Bank and from the shelf to the west at depths of at least 230 feet and perhaps more indicates that these now extinct land mammals once lived where the land is now covered by water. When the continental glaciers had their maximum extent, terminating at southern Long Island, Martha's Vineyard, Nantucket, and Georges Bank, the weight of the ice depressed the land so that the relation of the sea level to the land was not greatly different from what is is today. However, with the melting of the ice at the end of the glacial period and the retreat northward of the ice front, the land was relieved from the weight of the ice and rebounded so that sea level fell relative to the land. The maximum lowering of sea level appears to have occurred about 15,000 years ago and amounted to 440 feet. Since then, the sea level has gradually risen until the present condition was reached.

More detailed evidence, based chiefly on the carbon 14 content of peats, enables the rise in sea level during the past 4000 years to be considered on a regional basis. In Figure IX-3, curve A shows the trend of the change in sea level, relative to the land, at Cape Cod, Nantucket, and Long Island. Curve B shows the similar trend at positions north of Cape Cod, at Neponset and Plum Island. Curve C shows the trend at more southern positions, at Bermuda, North Carolina, Florida, and Louisiana. It is clear that the rise in sea level during the past 4000 years has been quite different in these three regions.

The change in sea level relative to the land may be attributed to two components: 1) the change due to the volume of water in the ocean, the *eustatic component,* and 2) the change due to the local warping of the earth, the *tectonic component.* The southern positions, on which curve C is based, come from a large area which may be assumed to be relatively stable and in which the tectonic component is negligible. If so, curve C may be considered to represent the eustatic component, which should be the same in all parts of the ocean. The differences between curves A or B and curve C represent their respective tectonic components. North of Cape Cod the land would appear to have subsided about 5 feet in the past 4000 years; at Cape Cod, Nantucket, and Long Island the subsidence has been about 15 feet in that time. Whether the southern group represent the true eustatic change or not it is clear that subsidence has been much greater in the Cape Cod region than to the northward.

References

Meany, C. C. Mean seal level—a basic engineering datum. Journal of the Coast and Geodetic Survey, 2, 34-37, 1949.

Hicks, S. O. On the classification and trends of long period sea level series. Shore and Beach, 20-24, April 1972.

Emery, K. O. and E. Uchupi. Western North Atlantic Ocean: Topography, Rocks, Structure, Water, Life and Sediments. American Association of Petroleum Geologists, Tulsa, 532 pp., 1972.

Redfield, A. C. Post-glacial change in sea level in the western North Atlantic Ocean. Science, 157, 687-692, 1967.

Chapter X

THE TIDE OFFSHORE

The coast may be divided into regions in which the local tides have a common origin in the deep ocean but in each of which the local tide is modified by its passage across the continental shelf and further changed by the topography of the local coast.

In the North Atlantic Ocean the tide appears to be engendered by a wave moving from the south. High water occurs earliest on the coast of North Africa, then progressively later off Spain, the British Isles, Iceland, Greenland, Newfoundland, and the eastern coast of the United States. The crest of high water moves in an anticlockwise direction around the North Atlantic basin completing the circuit in about the time required for it to reinforce the next tidal wave entering from the south. This is illustrated in Figure X-1 in which the cotidal lines show the positions where high water occurs at the same time. A very similar pattern is shown for the M_2 component. In both the position of the lines is conjectural except where they approach the coast, there being few observations available for the deep water.

On opposite sides of the North Atlantic, about 6 hours separate the times of high water. When it is high water on one side, it is low water on the other. It is deduced from this that there is an amphidromic point in midocean where there is no tide, and about which the tidal wave circulates. The amphidromic point appears to lie southeast of Newfoundland.

From these observations it is believed that the tide in the North Atlantic Ocean is an amphidromic system. Off the New England coast high water occurs about 12 hours after the moon's upper transit at Greenwich and the mean range is about 2.7 feet. These conclusions are supported by measurements made with a pressure recorder placed on the bottom near the edge of the continental shelf south of Martha's Vineyard in a depth of 276 feet which showed high water for the M_2 component to occur at 11.96 hours after the moon's transit at Greenwich and its range to be 2.76 feet.

The deep ocean may be considered to be bounded by the outer edge of the continental shelf where the depth decreases rather suddenly from several

The Tide Offshore

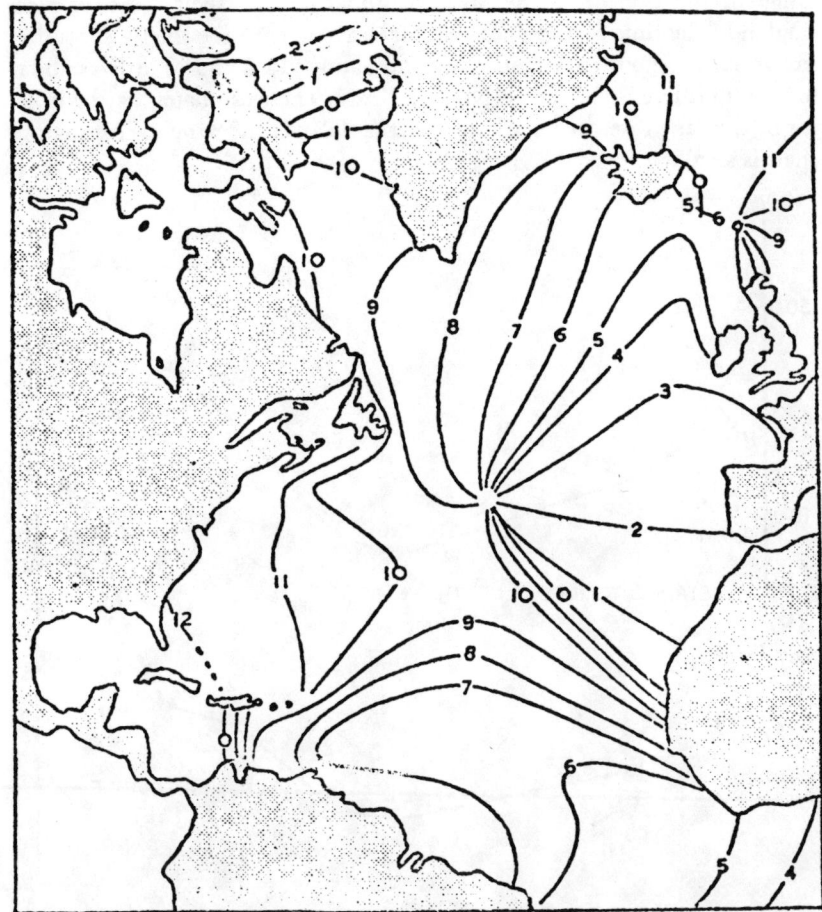

Fig. X-1. Cotidal lines of North Atlantic Ocean at successive lunar hours. (In part from Rept. Superintendent of U.S. Coast and Geodetic Survey for 1904. With permission.)

thousands of feet to about 300 feet. This boundary determines the character of the tidal system which occurs in the North Atlantic. Evidently, the dimensions of the enclosed basin are such that it has a natural period close to that of the M_2 constituent and, consequently, the ocean tide is chiefly semidiurnal.

At outlying positions on the coast west of Fire Island Inlet on Long Island the range of the tide varies with the width of the shelf, as shown in Figure X-2. Augmentation is apparently due to reflection from the coast of the tidal wave engendered in the deep oceans much as the tide in an embayment is augmented by reflection of the entering wave from the head of an embayment. Figure X-3 shows estimates and measurements of the mean range of the tide or of the M_2 constituent on the shelf south of New England. The time required for the tidal wave to cross the shelf is short, being about 0.3 hour at Atlantic City and Sandy Hook and as little as about 0.1 hour off Fire Island and at No Mans Land. These estimates may be in error but, in any case, they indicate that the delay of the wave in crossing the shelf is small.

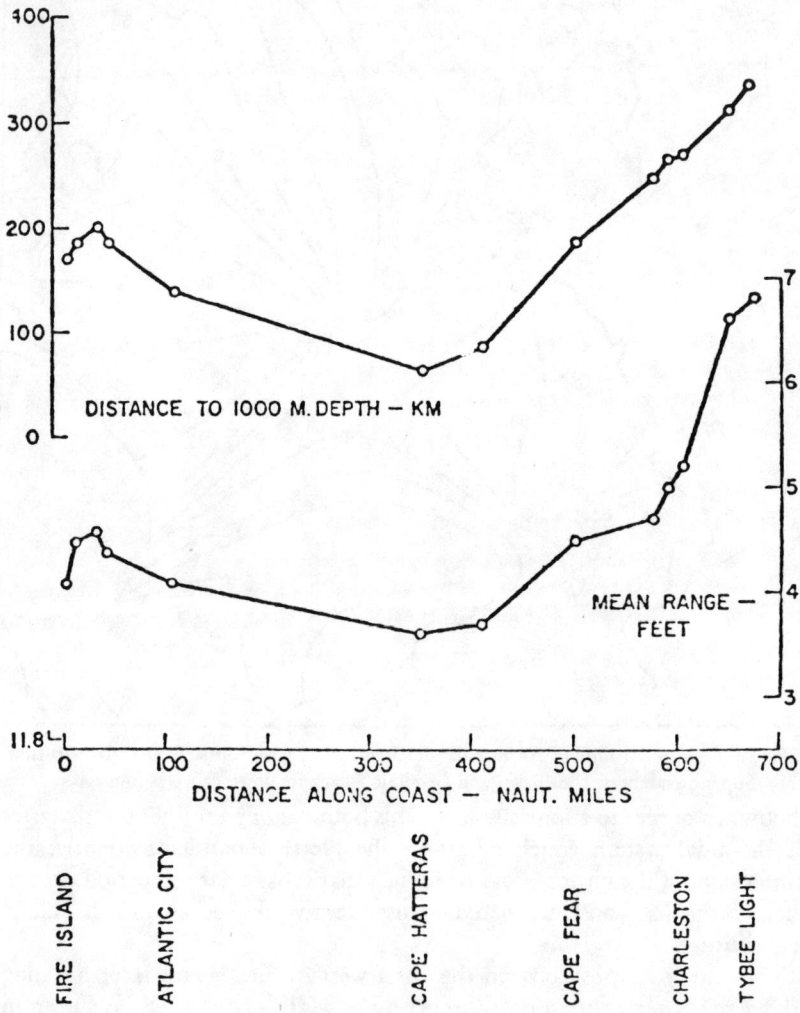

Fig. X-2. Width of continental shelf and mean range of tide at outlying positions west of Fire Island. Above: distance in kilometers from coast to 1000 meter depth contour. Below: mean range of tide at coastal positions. (From the *Journal of Marine Research*, vol. 17. With permission.)

The Tide Offshore

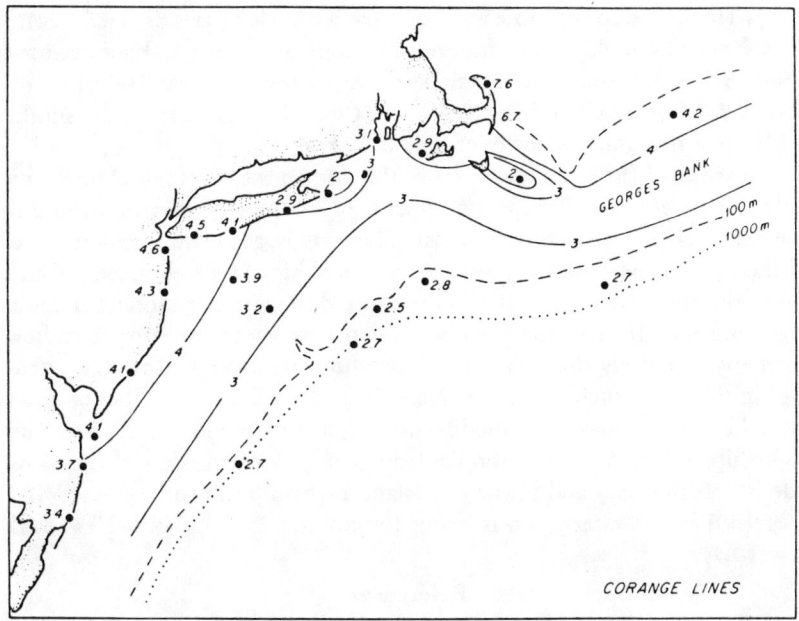

Fig. X-3. Mean range of tide, in feet, at coastal positions and on continental shelf and co-range lines on shelf west of Georges Bank.

East of Cape Cod the coast is separated from the deep ocean by shallows such as Georges Bank and by the deeper waters of the Gulf of Maine. Before reaching the coast of northern New England, the tidal wave engendered in the ocean must traverse a much greater distance than in the case of the southern coast. On the coast the mean range of tide is increased to 9 feet or more and the time of high water is greatly delayed. The tidal movement of the Gulf of Maine on which these observations depend is considered in the next chapter.

The tidal currents as observed from vessels anchored on the continental shelf are rotary in character. At a distance from the shore the current rose is a broad ellipse in which the direction changes each hour in a clockwise direction as illustrated by the currents at the Nantucket Lightship shown in Figure VII-1 (page 38). Very similar rotary currents occur on Georges Bank. These currents are strong as required by the large range of tide in the Gulf of Maine into which the tide water must flow across the relatively shoal water of the bank. Maximum velocities reach 1.5 to 2.0 knots and occur about 3 hours before and after high water on the coast of the Gulf. West of Nantucket the shelf is narrower, the range of tide is not augmented as much as in the Gulf of Maine, and the tidal currents on the shelf are much smaller. As measured from the U.S.S. *Fish,* when anchored 62 miles south by east from

Sandy Hook, the current rose was an ellipse with a long axis directed toward New York Lower Bay. The maximum current of about 0.1 knot occurred about 3 hours before or after high water. At the Barnegat Lightship anchored 7 nautical miles off the New Jersey Coast the tidal currents are similar but have a maximum velocity of only 0.05 knot.

Because of the rotary character of the tide on the continental shelf, it is only the onshore or offshore components of the motion that contribute to the rise or fall of the tide at the coast. These are large at intermediate stages of the tide but become small near the times of high and low water. At such times the components parallel to the shore dominate the motion. Thus at high tide the effect of the earth's rotation is to cause the current to flow northerly or easterly along the coast depending on its trend. However, openings in the coast, such as occur at New York, draw large quantities of water into the interior and may modify or reverse this tendency. At Scotland Lightship anchored 3 nautical miles from Sandy Hook the flow of the rising tide is northwesterly and at the Fire Island Lightship shown in Figure VII-2 (page 39) it is westerly, both being toward the opening into New York Lower Bay.

References

Redfield, A. C. The influence of the continental shelf on the tides of the Atlantic coast of the United States. Journal of Marine Research, 17, 432-458, 1958.

Haight, F. J. Coastal currents along the Atlantic coast of the United States. Coast and Geodetic Survey, U. S. Department of Commerce, Washington, Special Publication No. 230, 71 pp., 1942.

Chapter XI

THE WATERS OF NORTHERN NEW ENGLAND

The coast of New England north of Cape Cod opens on the Gulf of Maine where the behavior of the tide determines the tide in the numerous passages along the coast. The advance of the cotidal lines and the corange lines is indicated in Figure XI-1. While this diagram is somewhat speculative because of the limited data on the outer side of the Gulf, its general features have been confirmed by a mathematical model based on hydrodynamic principles.

The principal flow of tide water which causes the tide in the Gulf appears to enter from the east, producing a high water which occurs later at the western end. This flow is mainly through the East Channel which is much deeper than the water over the neighboring banks as shown in Figure XI-2. Recent measurements have shown that a tidal current having a maximum velocity of one knot occurs in the deep water of the East Channel. The character of the advance of the crest of high water within the Gulf is shown also by the wave curvature map in Figure XI-3 which is calculated from the changing rate of advance, and hence refraction, of a wave in water of varying depth. The high velocities of the tidal currents on Georges Bank suggests that they may contribute to the tides in the Gulf, but because of the shallowness of the water their volume is small. Their effects are probably not very large except possibly along the eastern coasts of Cape Cod and Nantucket.

The relation of the cotidal and corange lines, shown in Figure XI-1, which cross at a large angle, suggest that the tidal movement in the Gulf of Maine is a degenerate amphidromic system in which the amphidromic point, if it existed, would lie southwest of Cape Cod. The Gulf is too deep for its size to develop a full amphidromic system.

The coast of the state of Maine consists of a series of headlands separating embayments of various length formed by the drowning of pre-existing valleys by the post-glacial rise in sea level. *Penobscot Bay* is the largest of these, extending for 47 nautical miles from the outlying island of Monhegan to its head at Fort Point. Nomographic analysis, taking Fort

68   The Tides of the Waters of New England & New York

Point where $A = 2.575$ feet to be the position of reflection, indicates that the tide at Penobscot Bay is a reflected co-oscillation, R being 1 and $\mu = 1$. When these values are introduced into the theoretical equations, Chapter II, the mean ranges and the times of high water and slack water calculated agree with those predicted about as precisely as the predictions are given (Figure XI-4).

The *Penobscot River*, which enters Penobscot Bay at Fort Point, is blocked by a dam at Bangor which stops the advance of the tide. Below Bangor the tide in the river appears to be a co-oscillation between the enter-

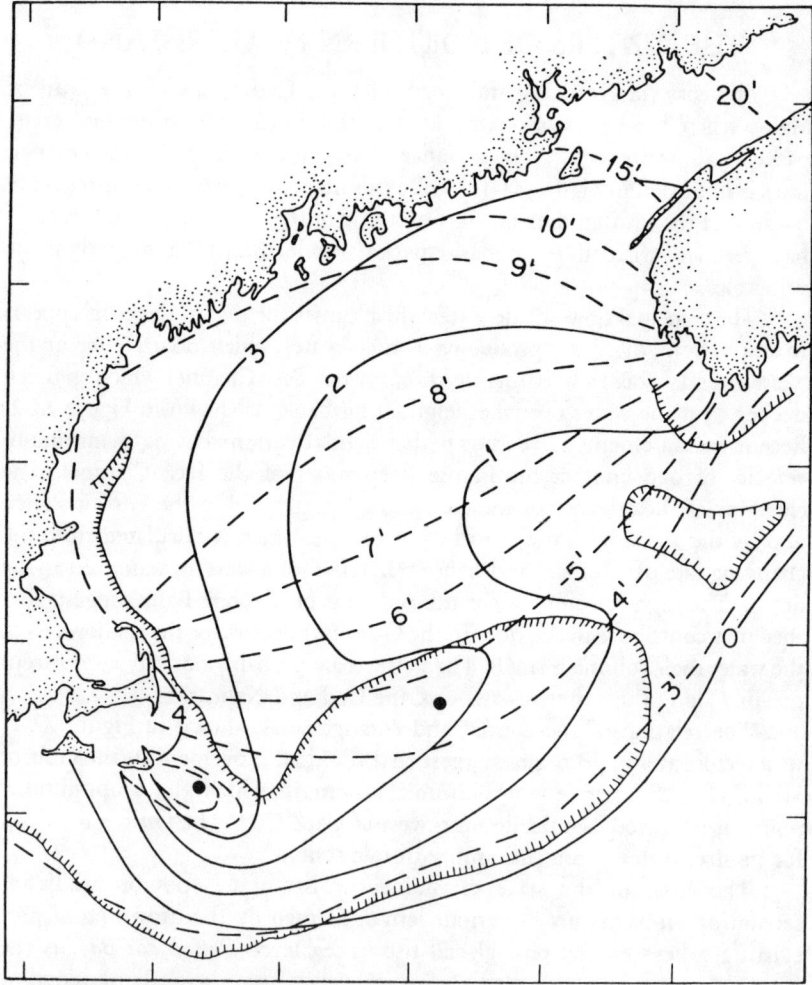

Fig. XI-1. The Gulf of Maine — cotidal and corange lines. Solid lines: Cotidal lines numbered in lunar hours relative to moon's transit at Greenwich. Broken lines: corange lines numbered in feet.

The Waters of Northern New England

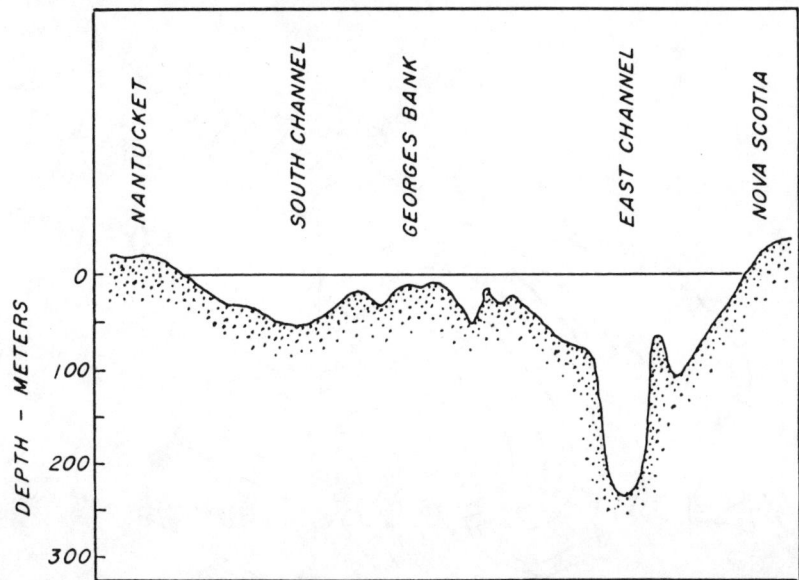

Fig. XI-2. Depth of sill separating the basin of the Gulf of Maine from the deep ocean.

ing wave and its reflection from the dam. Nomographic analysis, taking Bangor where $A = 3.275$ feet to be the position of reflection, indicates that $R = 1$ and $\mu = 1.5$. When these values are introduced into the theoretical equation, the mean ranges and the times of high water agree closely with those predicted (Figure XI-5).

The many other embayments on the Maine Coast are similar to Penobscot Bay and the lower reach of the river except that they are relatively short and the tide undergoes a relatively small change in phase during its ascent. In general, the range is increased by not more than about one-half a foot and the time of high water is delayed by not more that 0.2 hour. Tentatively, it may be concluded that the tide in these embayments is also a reflected co-oscillation.

The drowning of this coast by the post-glacial rise in sea level has formed a number of straits behind or between islands, such as *Eggemoggin Reach, Fox Island Thorofare, Deer Island Thorofare, and Moosabec Reach*. The current tables give no data for the currents in these straits. These passages connect bodies of water in which the tides are so similar that strong hydraulic currents do not develop, except where local obstruction may produce local hydraulic effects. The Coast Pilot does not mention the currents in Eggemoggin Reach and states that they are not strong in the Fox Island and Deer Island thorofares. In Moosabec Reach it states that the currents have "considerable velocity." In the Fox Island Thorofare it says that the

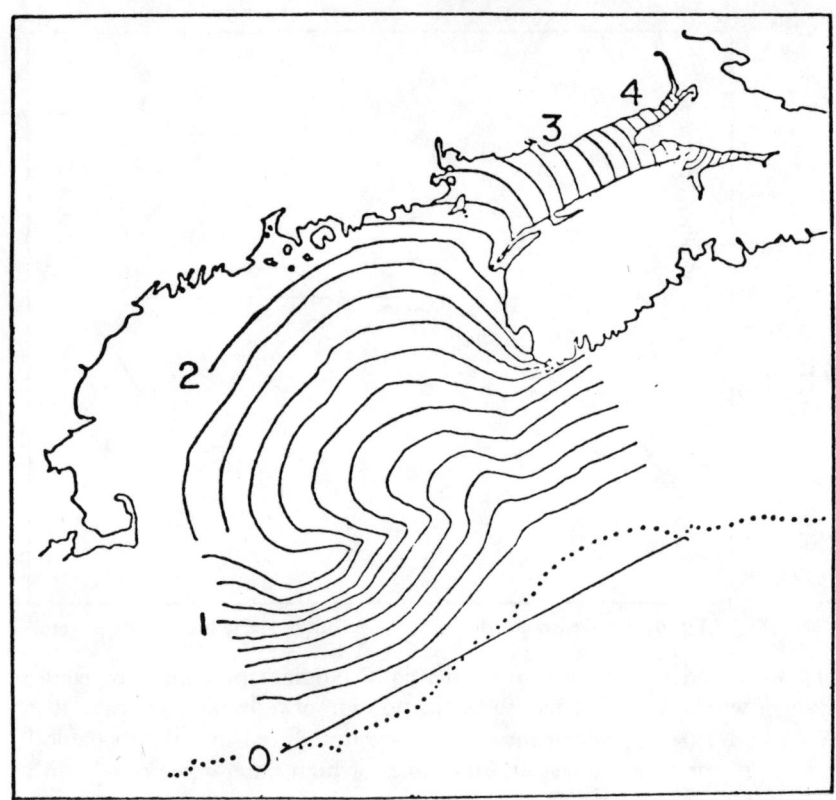

Fig. XI-3. Refraction of a linear wave front entering the Gulf of Maine. Contour interval 10 minutes. Numbers indicate time of advance in hours. (After Duff in J. Fish. Res. Bd., Canada, Vol. 27. With permission.)

tidal currents set in from both ends and meet at Iron Point in the middle of the thorofare.

The *Bay of Fundy* is the largest tributary to the Gulf of Maine. It extends between New Brunswick and Nova Scotia for 120 nautical miles northeastward from Grand Manan. Nomographic analysis taking the tide at Hopewell Cape in the Chegnecto Channel, where the mean range is 33.2 feet, to represent the position at which the entering tidal wave is reflected indicates that the tide is a reflected co-oscillation in which $R = 1$ and $\mu = 1$. Introducing these values into the theoretical equations (Chapter II) shows that the predicted values for the mean range, and the intervals of high water and slack water agree satisfactorily with their calculated values at most positions (Figure XI-6).

The *Minas Basin* is considered to be a tributary of the Bay of Fundy, entering it at about the position of Ile Haute. This treatment is required

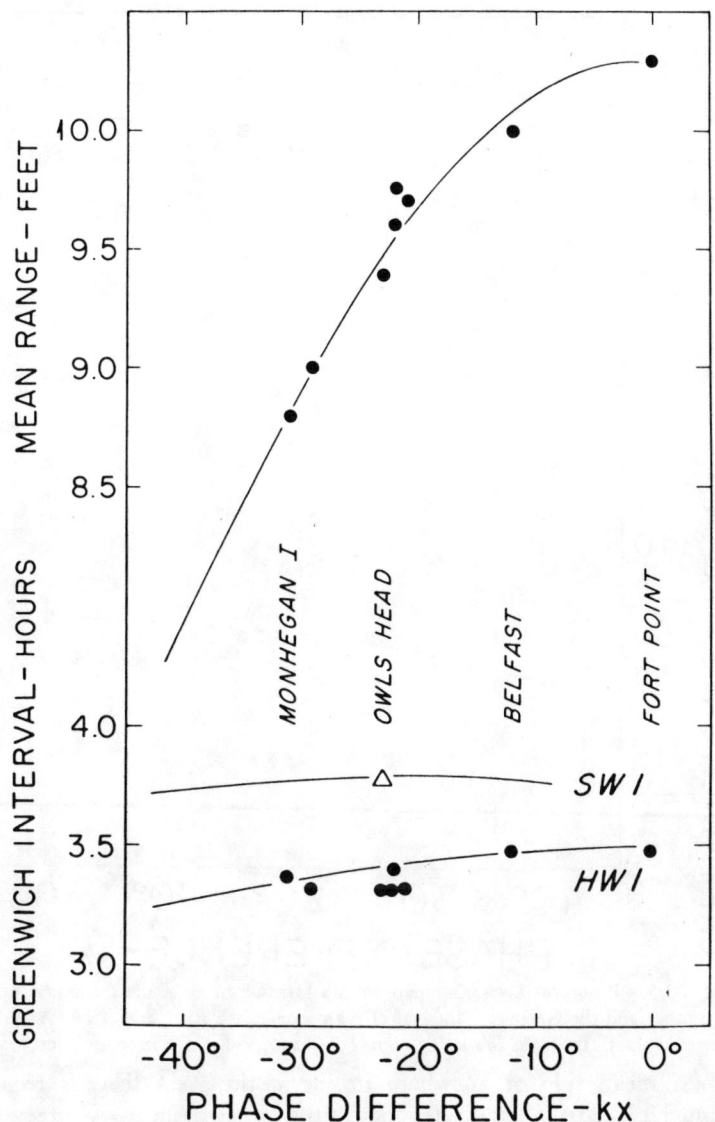

Fig. XI-4. Penobscot Bay. Comparison as a function of kx of the theoretical values of mean range and the Greenwich intervals of high and slack water when $R = 1$ and $\mu = 1$ with their predicted values. (From the *Journal of Marine Research*, Vol. 36. With permission.)

because of the greater attenuation which must be assigned to the tide in Minas Basin than to other parts of the Bay of Fundy. The tide in the Minas Basin may be considered to be a reflected co-oscillation in which the wave entering from the Bay of Fundy is reflected at Burntcoat Head where the

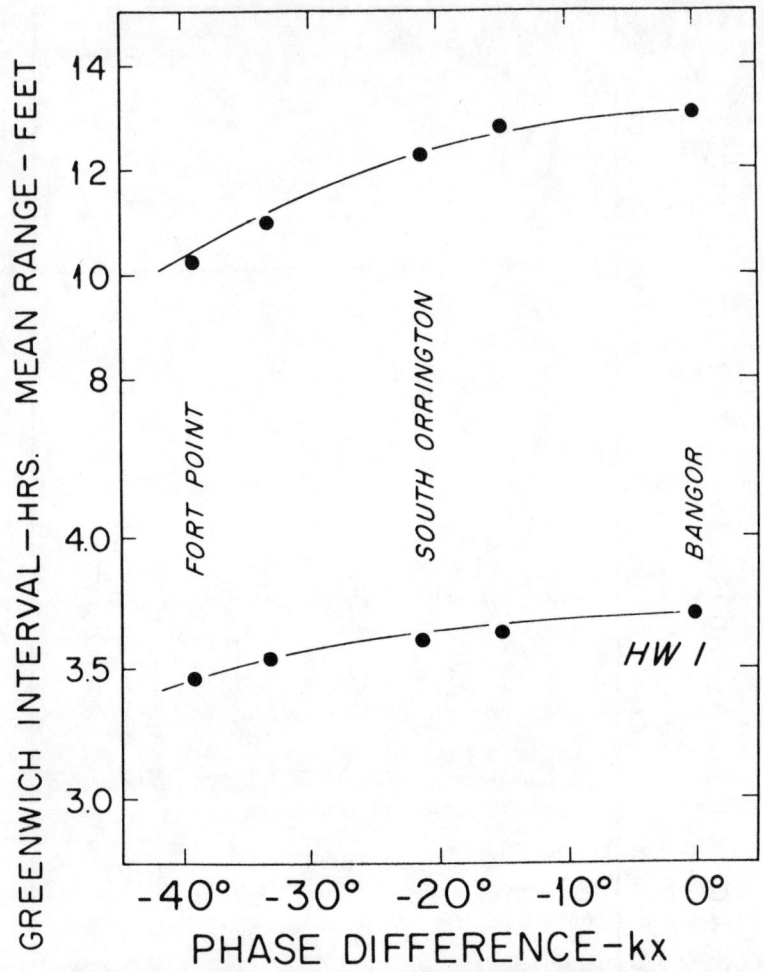

Fig. XI-5. Penobscot River. Comparison as a function of *kx* of the theoretical values of mean range and the Greenwich interval of high water when $R = 1$ and $\mu = 1.5$ with their predicted values. (From the *Journal of Marine Research*, Vol. 36. With permission.)

highest mean tide of anywhere in the world, 38.4 feet, is recorded. Nomographic analysis indicates that in the Minas Basin $A = 9.6$ feet, $R = 1$ and $\mu = 3$ (Figure XI-7).

The great range in tide at the head of the Bay of Fundy is responsible for the tidal bores which occur in the shallow rivers which enter there and for the reversing falls which occur in the St. John River. On reaching Burntcoat Head in the Minas Basin the mean range in the tide of the ocean amplified about 14-fold, from 2.75 to 38.4 feet. This is much greater than is to be expected in the case of a reflected co-oscillation. It is permissible to think this

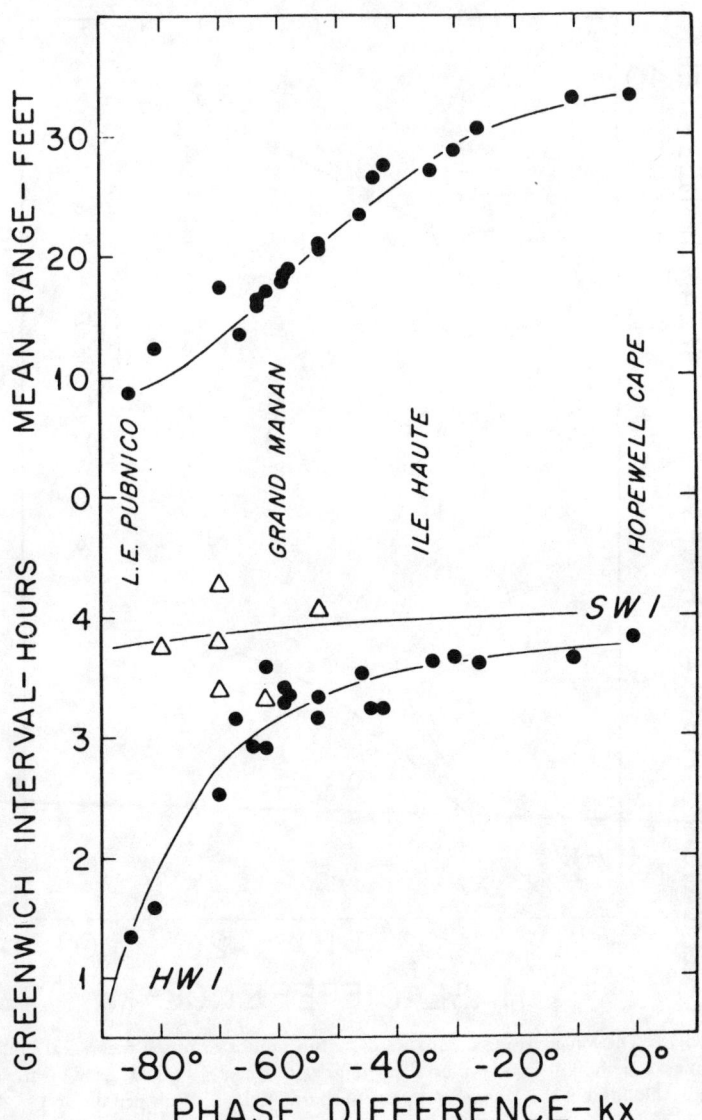

Fig. XI-6. The Bay of Fundy. Comparison as a function of *kx* of the theoretical values of mean range and the Greenwich intervals of high water and slack water when $R = 1$ and $\mu = 1$ with their predicted values. (From the *Journal of Marine Research*, Vol. 36. With permission.)

to be the result of a series of steps in amplification in each of which the range is augmented by the passing of the wave from a large body of water to a smaller embayment. Thus the ocean tide is increased from a mean range of 2.75 feet to one of about 9.5 feet in the Gulf of Maine, an augmentation of

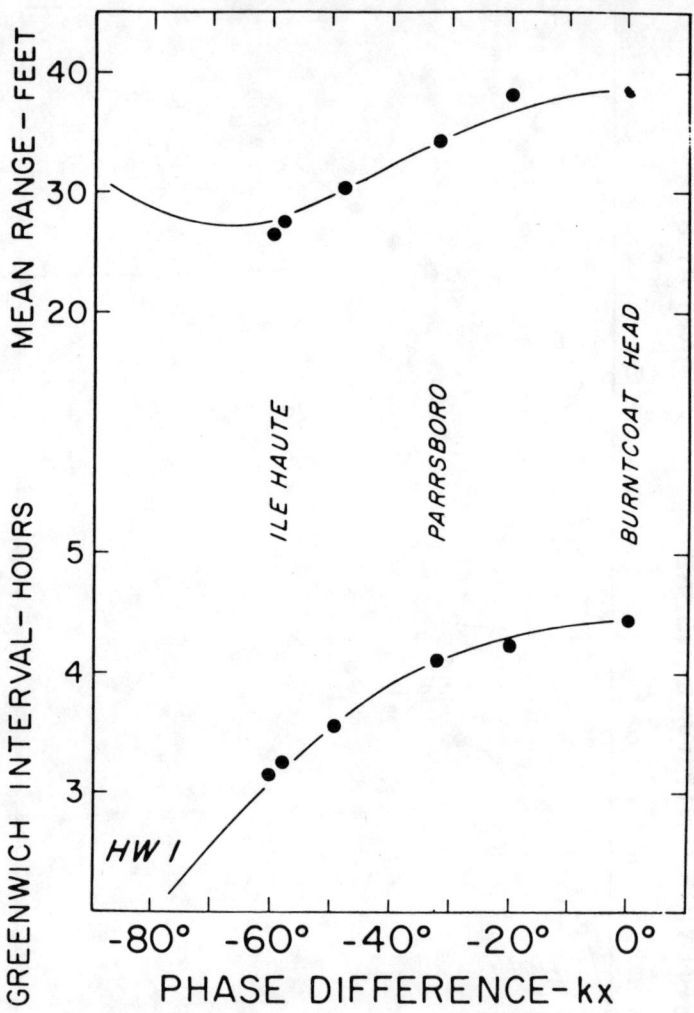

Fig. XI-7. The Minas Basin. Comparison as a function of kx of the theoretical values of mean range and the Greenwich interval of high water when $R = 1$ and $\mu = 3$ with their predicted values. (From the *Journal of Marine Research*, Vol. 36. With permission.)

about 3.5-fold. The 9.5 foot tide of the Gulf of Maine, is increased to 27.5 feet at Ile Haute in the Bay of Fundy, an augmentation of about 2.9-fold. The tide at Ile Haute is increased to 38.4 feet at Burntcoat Head in the Minas Basin, an augmentation of about 1.4-fold. The product of these three successive augmentations is about 14-fold.

The effect of the tide in the Bay of Fundy on that in the Gulf of Maine is evident from the mean range of tide predicted for positions along the coast of Maine. It increases eastward to more that 10 feet near Mt. Desert Island and continues to increase until values in excess of 30 feet are reached toward the head of the Bay of Fundy. Geographically, the Bay of Fundy is considered to lie northeast of the line between Grand Manan and Brier Island. The mean range of tide at Grand Manan is given as 17 feet so that evidently the tide in the Bay of Fundy has an effect far beyond this limit. Nomographic analysis indicates that the difference in phase between the position of reflection at Hopewell Cape in the Chegnecto Channel and Grand Manan is about 60° and that the reflected co-oscillation of the Bay of Fundy dominates the tide to beyond Lower East Pubnico where the phase difference is 85°. An analysis by Duff using hydrodynamic principles also extends the Bay of Fundy system beyond Lower East Pubnico to a position near the outer edge of the continental shelf. Recently, a numerical model based on hydrodynamic principles has been formulated on the assumption that the Bay of Fundy and the Gulf of Maine together act as one tidal system in close resonance with the semidiurnal tides. This model reproduces closely the predicted tide at position along the coast. These theoretical approaches are not inconsistent.

At the western end of the Gulf of Maine, on the shores of New Hampshire and eastern Massachusetts, quantities of unconsolidated glacial deposits have been left by the retreat of ice. The erosion of these deposits has supplied large quantities of sand to form barrier islands and beaches which enclose shallow embayments, as behind Plum Island and at Barnstable and on a smaller scale elsewhere. At Barnstable the mean range of tide at Slough Point where the large salt marsh creeks enter the open water of the harbor is very nearly the same as at Beach Point where the tide enters the harbor. High water is delayed only 0.1 hour between these positions, a distance of about 2.9 nautical miles. Because the harbor is nearly empty at low water, the depth is continually increasing as the tide rises and the theory for the tide in ideal embayments cannot be applied. The rise and fall of the tide in this enclosure may be considered to be a response to the head developed as the water rises and falls with the tide in Cape Cod Bay; that is, it is essentially hydraulic in nature. A similar explanation may be applied probably to the tide in the enclosure behind Plum Island and to other embayments along the western coast of the Gulf of Maine.

In *Boston Harbor*, which is deeper than the enclosures at Barnstable and behind Plum Island, the mean range increases about 0.5 foot between Boston Light at the entrance and positions along the inner shore. Probably its tide may be considered to be in reflected co-oscillation with the tide in the Gulf of Maine. The National Oceanic and Atmospheric Administration issues Tidal Current Charts of Boston Harbor which show the strength and direction of the current at hourly intervals after high or low water at Boston.

In the *Piscataqua River* the presence of a number of large embayments near its mouth causes the theoretical analysis of the tides to be impossible.

On the eastern shores of Cape Cod and Nantucket Island which face the open water of the Gulf of Maine and of Nantucket Shoals, the range of tide falls progressively and the predicted time of high water becomes earlier in passing from the Gulf to the continental shelf. At Cape Cod Lighthouse the mean range is 7.6 feet and high water is predicted at 3.92 hours after the moon's transit at Greenwich. At Tom Nevers Head on the southeast corner of Nantucket the mean range is only 1.2 feet and high water occurs 2.97 hours after the moon's transit. The very low range at Tom Nevers Head suggests that the range of tide along this stretch of coast is due to the interference of waves entering from the ocean and those moving out of the Gulf of Maine in the opposite direction. The situation is similar to that in Vineyard and Nantucket sounds, described in Chapter III, except that here $R = 3$ and $\mu = 3$. These facts suggest that the exchange of tidewater across the banks, at least in the region of the South Channel, may have some influence on the tide in the Gulf of Maine, but the magnitude of its effect cannot be evaluated.

While the principal movements of the tide in the Gulf of Maine, on which the tides along its coast depend, are reasonably clear, it must be admitted that in detail the theoretical interpretation is uncertain. Confirmation of the theory must wait until more measurements are made of the tide in the deep water of the Gulf and on the offlying banks.

References

Redfield, A. C. Interference phenomena in the tides of the Woods Hole region. Journal of Marine Research, 12, 121-140, 1953.

Redfield, A. C. The tide in coastal waters. Journal of Marine Research, 36, 255-294, 1978.

Duff, G. F. D. Tidal resonance and tidal barriers in the Bay of Fundy system. Journal of the Fisheries Research Board of Canada, 27, 1701-1728, 1970.

Greenberg, D. A. A numerical model investigation of tidal phenomena in the Bay of Fundy and Gulf of Maine. Marine Geodesy, 2, 161-187, 1979.

Chapter XII

THE WATERS OF SOUTHERN NEW ENGLAND

The southern coast of New England is separated from the continental shelf by a series of islands, Long Island, Martha's Vineyard, and Nantucket. The tidal wave generated in the deep ocean reaches the outer shore of these islands after a very short period and is augmented by reflection to a mean range of 4.1 feet at the Fire Island Breakwater and 3 feet at No Mans Land (Figure X-3, page 65).

On the south coast of Long Island a number of shallow bays occur separated from the continental shelf by barrier beaches or islands. The largest of these is *Great South Bay* which opens through Fire Island Inlet. The tide in the inlet appears to be hydraulic in nature, the mean range falling from 4.1 feet at the outer end of the breakwater to 2.6 feet at Democrat Point and to 0.7 foot north of Fire Island Light where the inlet opens into the bay. Within the bay the tide appears to be a reflected co-oscillation, between the tide entering through the inlet and its reflection from the head of the bay at Bellport. Nomographic analysis, taking Bellport where $A = 0.2$ foot as the position of reflection, indicates that $R = 1$ and $\mu = 3$. The analysis is not very exact because the precision with which the mean ranges at positions in the bay are given is not much greater than their variation from one another. However, when these values of R and μ are introduced into the theoretical equations given in Chapter II, the calculated values of the mean range and the interval of high water agree reasonably well with the predicted values at the recorded positions (Figure XII-1).

Moriches Bay and *Shinnecock Bay* now open through inlets onto the continental shelf. Sufficient information is not available to warrant exact analysis of their present day tides. Formerly, in 1907, the beach was continuous from east of Shinnecock Bay to Fire Island Inlet and inlets into these bays did not exist. The bay water was much fresher than at present and the tidal conditions must have been quite different.

Hempstead Bay is part of a strait connecting Great South Bay and the continental shelf at Jones Inlet. Nomographic analysis indicates that the posi-

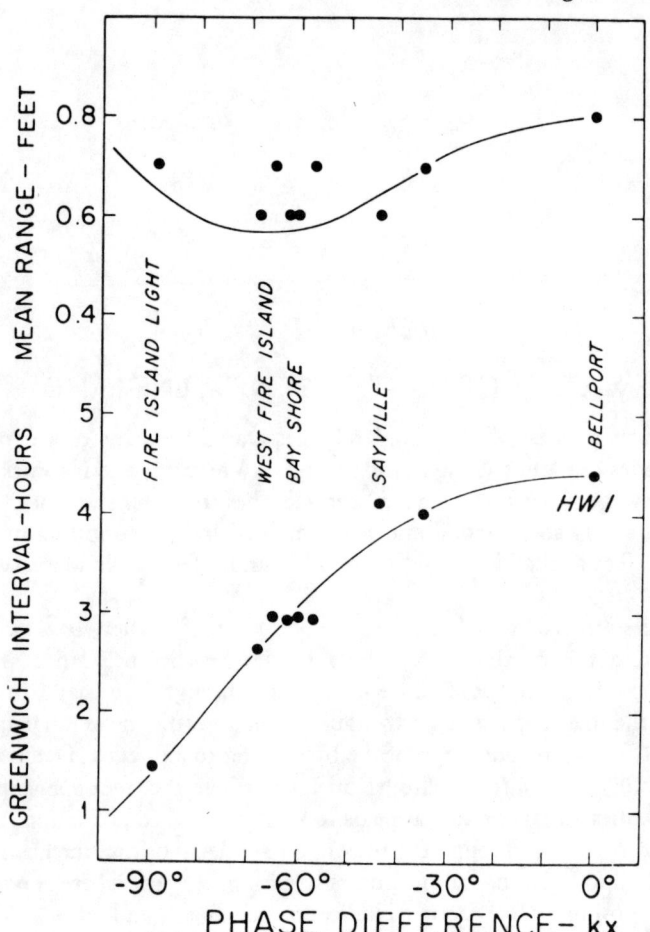

Fig. XII-1. Great South Bay. Comparison as a function of kx of the theoretical values of mean range and the Greenwich high water interval when $R = 1$ and $\mu = 3$ with their predicted values. (From the *Journal of Marine Research*, Vol. 36. With permission.)

tion of phase equality is near Amityville where $A = 0.5$ foot and that $R = 0.2$ and $\mu = 7$. The large value of the μ is probably related to the small depth in Hempstead Bay which for the most part is less than 2 feet. When these values of R and μ are introduced into the theoretical equations, Chapter II, the theoretical values for the mean range and the interval of high water agrees well with those predicted for positions along the strait (Figure XII-2).

The principal access of the ocean tide to the inner coast of southern New England is between Montauk Point on the eastern end of Long Island

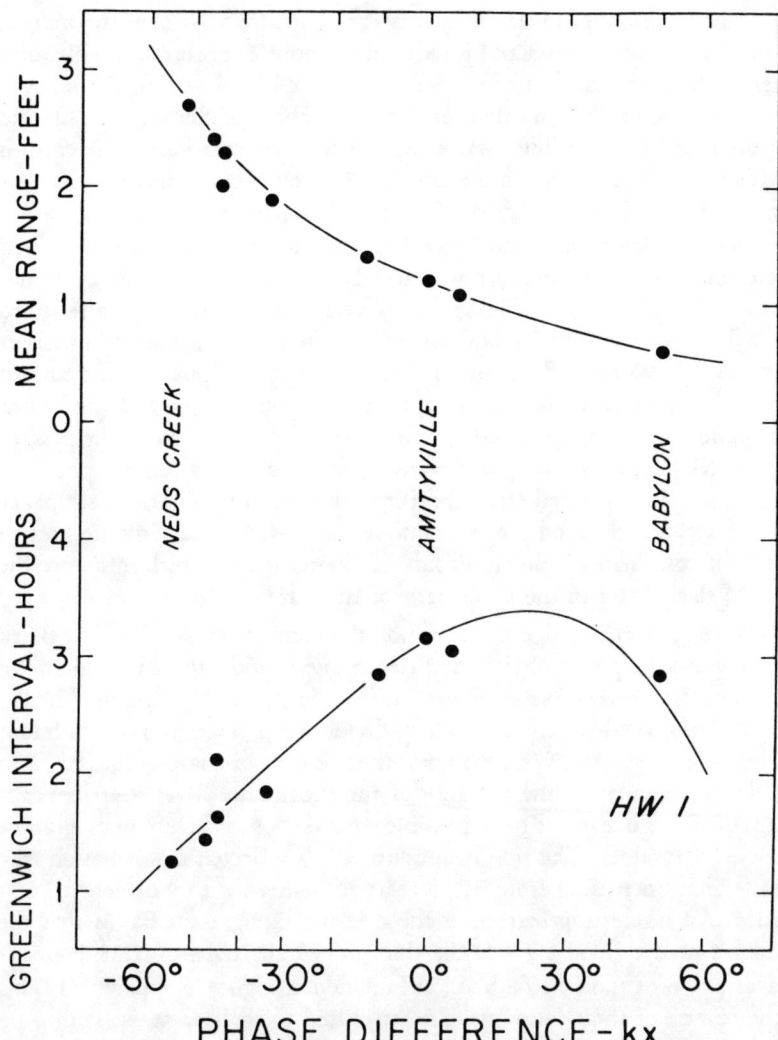

Fig. XII-2. Hempstead Bay. Comparison as a function of kx of the theoretical values of mean range and the Greenwich high water interval when $R = 0.2$ and $\mu = 7$ with their predicted values. (From the *Journal of Marine Research*, Vol. 36. With permission.)

and Gay Head on the western end of Martha's Vineyard where the tide in a number of embayments and straits originates. The largest of these is the *Long Island Sound system*, discussed in Chapter IV, where it is shown that it may be treated either as a strait or a reflected co-oscillation having a position of phase equality or of reflection near Glen Cove. In either case, $R = 1$ and $\mu = 1$. As the result of interference of the waves moving in opposite directions, the mean range of tide is increased to more than 7 feet at the western end of Long Island Sound and is reduced to about 2 feet off Montauk Point where a node occurs. East of the Race, in Block Island Sound, the range is greater on the north shore than on the south shore of the sound. This and the predicted times of high water suggest that the rotation of the earth is having a significant effect on the tide. The movement appears to be that of a degenerate amphidromic system (Figure VII-7, page 44).

Peconic Bay opens onto Block Island Sound via Gardiners Bay at its western end through two narrow channels on either side of Shelter Island. The tide in Peconic Bay is shown by nomographic analysis to be a reflected co-oscillation in which the position of reflection is near South Jamesport where $A = 0.675$ foot, $R = 1$ and $\mu = 4$. When these values are introduced into the theoretical equations, Chapter II, the mean range and interval of high water agree closely with those predicted for positions along the passage (Figure XII-3). Slack water, however, is predicted to occur about one hour earlier than is calculated from the theoretical equations. This discrepancy may be accounted for on the assumption that hydraulic currents are present in the narrow channels on either side of Shelter Island which influence the time of slack water in the inner parts of the system.

Narragansett Bay opens directly on the continental shelf east of Block Island. From its position, size, and depth one would expect the tide to be much as it is in Buzzards Bay. However, its topography is complicated by the presence of several large islands which divide the passage into two different channels, the East and West Passages, the presence of a large tributary from Mt. Hope Bay and of the entrance of the Providence River near its head. Consequently, it has not been possible to make a satisfactory nomographic analysis of its tides. The tide is undoubtedly a reflected co-oscillation, the mean range increasing from 3.5 feet at the entrance to 4.6 feet at Providence. An interesting feature of the tide in Narragansett Bay is that the duration of the flood exceeds the duration of the ebbs by 1.28 hours at Beavertail Point and by 1.48 hours at Providence (Figure VI-4 page 34). This is the reverse of the relation usually attributed to shallow water. It may be accounted for by the timing of the combined harmonics which is such as to prolong the period of the rising tide, but this timing is not explained.

Buzzards Bay opens on the continental shelf off Cuttyhunk Island. The mean range at the entrance is 3.4 feet and increases to about 4.1 feet at the head of the bay after a delay of about 0.5 hour. Nomographic analysis indicated that the tide is a reflected co-oscillation. Taking the tide at Great Hill where $A = 1.025$ feet to represent that at the position of reflection, $R = 1$ and $\mu = 3$. When these values are introduced into the theoretical equations in Chapter II, the theoretical values of mean range and high water interval agree closely with those predicted for positions on the bay. Slack

The Waters of Southern New England

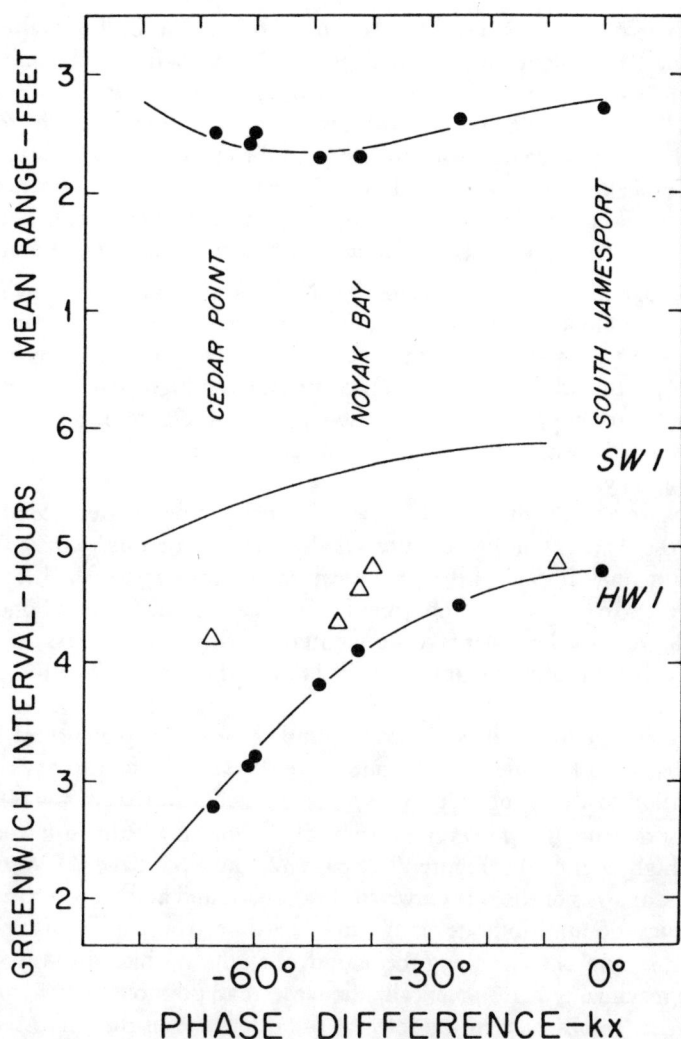

Fig. XII-3. Peconic Bay. Comparison as a function of *kx* of the theoretical values of mean range and the Greenwich intervals of high water and slack water when $R = 1$ and $\mu = 4$ with their predicted values. (From the *Journal of Marine Research*, Vol. 36. With permission.)

water occurs a little earlier than expected in theory but it is uncertain whether this indicates the presence of hydraulic effects or is due to small errors in the predictions (Figure XII-4). Like Narragansett Bay, the duration of the rising tide is longer than that of the falling tide and is similarily explained (Figure VI-3, page 33).

Vineyard Sound and *Nantucket Sound* together form a strait which is discussed in Chapter III. The tide enters from the continental shelf off Gay

Head where the mean range is 2.9 feet and also from the Gulf of Maine off Monomoy Point where the mean range is 3.7 feet. Within the strait the characteristics of the tide are determined by the interference of the waves entering from these two ends and moving in opposite directions. The result is that the mean range decreases greatly from the entrance at Gay Head to a node near Falmouth where it is only 1.3 feet and then increases toward Monomoy Point. The high water interval is 3.85 hours greater at Falmouth than at Gay Head, then increases more slowly toward Monomoy Point.

Nomographic analysis indicates that if the position of phase equality is taken as near Monomoy Point, where $A = 0.578$ foot, $R = 2.2$ and μ is 3.4, and if these values are introduced into the theoretical equations given in Chapter II, the mean range and the intervals of high and slack water calculated agree approximately with those predicted (Figure III-3, page 16). There is no evidence that hydraulic currents occur in this strait except perhaps locally.

Because of the reduction of the semidiurnal tide by interference in this system, the character of the tide curves is altered and it becomes a mixed tide which is mainly diurnal as the node near Falmouth is approached (Figure I-2, page 4). In a tributary embayment east of Falmouth, Bournes Pond, in which the very shallow inlet favors the entrance of the spring tides, the tide curve becomes practically that of a diurnal tide at that time (Figure I-3, page 4).

In the Vineyard and Nantucket Sounds system the presence of harmonics has a marked effect on the tide curve. West of Falmouth they cause the duration of the rising tide to be much shorter than that of the falling tide. East of Falmouth the reverse is true. The change is at Falmouth where a double highwater occurs (Figure VI-2, page 32, and VI-5, page 35). A recent harmonic analysis of the tide curves in Vineyard Sound and the western part of Nantucket Sound indicates that while the time at which the M_2 constituent is maximal changes along the Sound, the time at which the M_4 constituent is maximal remains practically the same at all positions. This accounts for the change observed in the form of the tide curve in the strait, but its physical cause is not clear.

The tide of the continental shelf has access to Nantucket Sound also through the Muskeget Channel. This tidal flow appears to have no significant effect on that in the Vineyard and Nantucket Sounds system except in the southern part of Nantucket Sound. There is a node near Wasque Point where the mean range is reduced to 1.1 feet. North of Muskeget Island where the mean range is 2 feet and the water is very shallow, strong winds have a great effect on the depth of the water.

The National Oceanic and Atmospheric Administration has issued charts which show the direction and strength of the currents in Nantucket

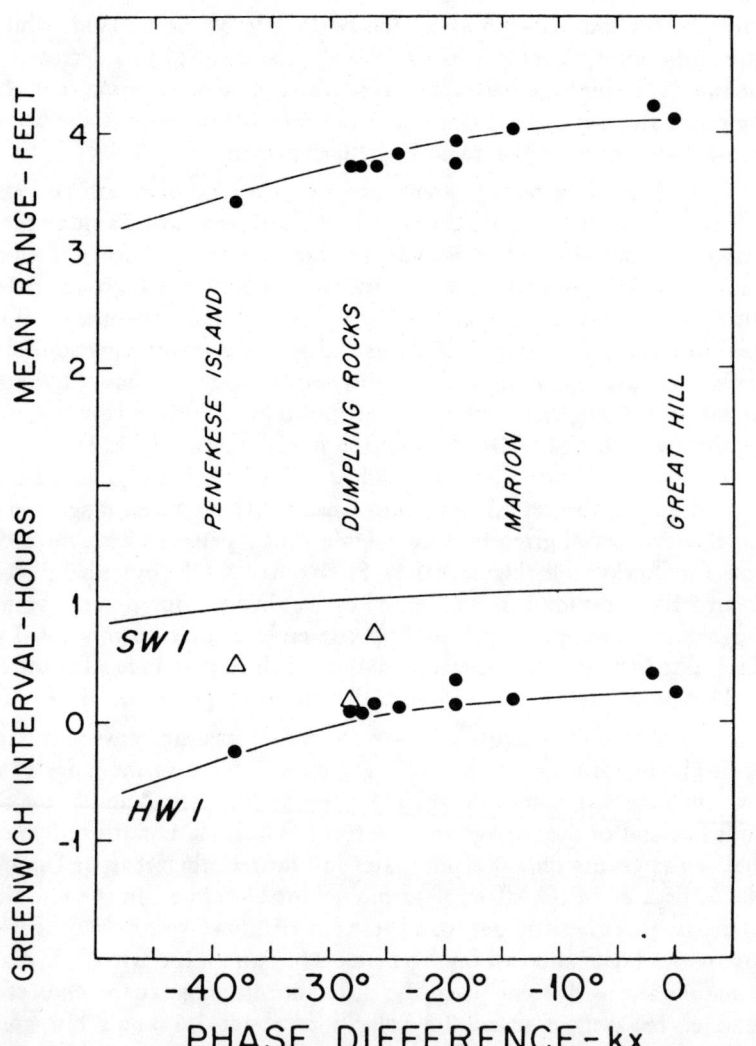

Fig. XII-4. Buzzards Bay. Comparison as a function of kx of the theoretical values of mean range and the Greenwich intervals of high and slack water when $R = 1$ and $\mu = 3$ with their predicted values. (From the *Journal of Marine Research*, Vol. 36. With permission.)

and Vineyard sounds, and of Narragansett Bay at hourly intervals relative to slack ebb or slack flood in Pollock Rip Channel.

The *passages between the Elizabeth Islands* and between them and the mainland at Woods Hole are locally known as *Holes*. They connect Buzzards Bay with Vineyard Sound. Because the tide in Buzzards Bay is a reflected co-oscillation, the range of tide and the interval of high water increase somewhat to the eastward along the north side of these islands while in Vineyard Sound, it being a part of a strait, the range of tide decreases and the interval of high water increases greatly to the eastward. As a result, both the range and the time of high water differ at either end of these passages giving rise to heads which cause hydraulic currents.

Quicks Hole, a passage about one nautical mile in length and about half this in width, separates the islands of Nashawena and Pasque near the western end of the Elizabeth Islands. The mean range is 3.5 feet at the north side of Quicks Hole and 2.5 feet at the south side, while high water occurs almost simultaneously everywhere. In the middle the current flows northward into Buzzards Bay for 6.8 hours and in the opposite direction for 5.6 hours. the average minimum velocity of the tide is about 2.5 knots. Nomographic analysis, assuming the position of phase equality to lie somewhere in Buzzards Bay beyond the north end of Quicks Hole, where $A = 0.6$ foot, indicates that $R = 2$ and $\mu = 2$. when these values are introduced into the theoretical equations, Chapter II, the mean range and the high water interval given by theory agree almost perfectly with the predictions for Quicks Hole (Figure XII-5). Slack water, on the other hand, occurs about 2 hours earlier than is indicated by the theory of interference between progressive waves entering from opposite ends. It may be concluded that this is due to hydraulic currents caused by the difference in head at the ends of the passage, and which dominate the flow through Quicks Hole.

The *Woods Hole passage* is the easternmost opening between Buzzards Bay and Vineyard Sound and has a greater difference in the range of tide than the others. The mean range is 3.6 feet at Uncatena Island at the Buzzards Bay end of the passage and 1.8 feet in Great Harbor at its other end. High water occurs only 0.2 hour later in Great Harbor than at Uncatena Island, because of the effect of harmonics in the former. In the middle of the passage, off Devils Foot Island, the current flows west toward Buzzards Bay for 5.3 hours and east for 7.1 hours. The current is obviously hydraulic in nature and is due chiefly to the difference in range at the ends of the passage. Its average maximum velocity is about 3.5 knots but greater velocities occur at the time of spring tides. At such times on a still night the sound of the turbulent flow through the Hole can be heard in the village of Woods Hole. The head which develops between Buzzards Bay and Great Harbor has a value of about one foot. The factors which produce this head and its variation during the tidal cycle are indicated in Figure XII-6. The topography of the Woods Hole Passage is too complicated to permit a meaningful nomographic analysis.

The Cape Cod Canal connects Cape Cod Bay on the northern coast of New England with Buzzards Bay on the southern coast. At the head of Buzzards Bay the mean range is about 4.1 feet while in Cape Cod Bay it is about

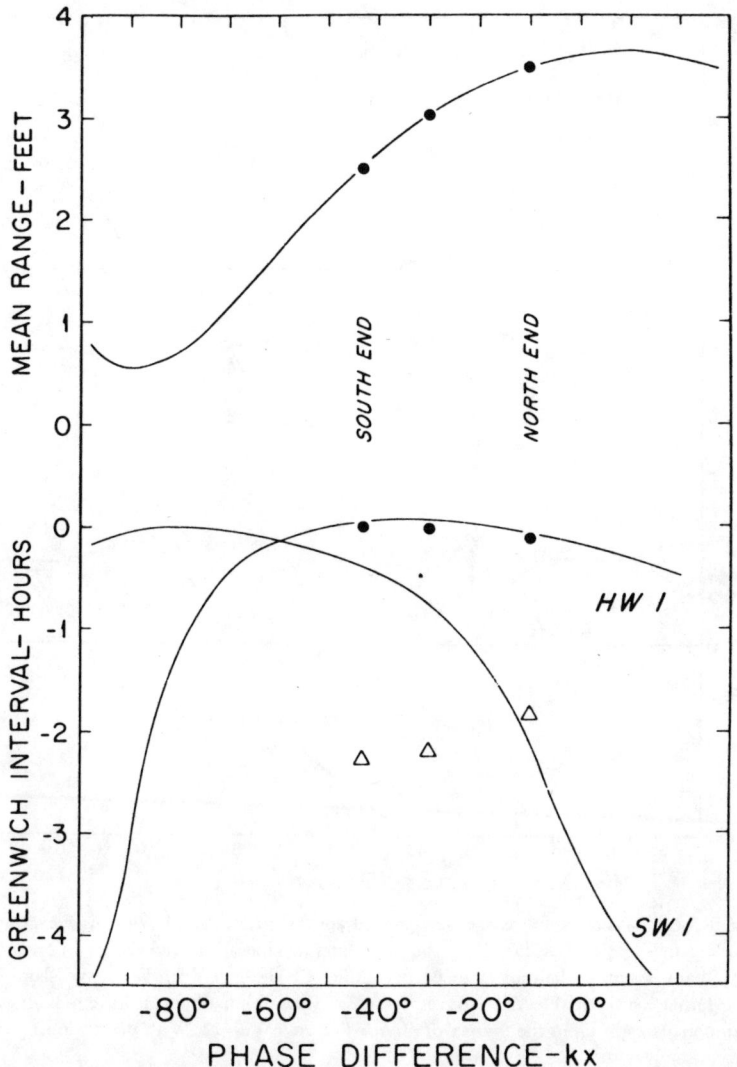

Fig. XII-5. Quicks Hole. Comparison as a function of kx of the theoretical values of mean range and the Greenwich intervals of high and slack water when $R = 2$ and $\mu = 2$ with their predicted values. (From the *Journal of Marine Research*, Vol. 36. With permission.)

9.5 feet and high water occurs about 3.25 hours later. The differences in range and time of high water suggest that strong hydraulic currents occur in the Cape Cod Canal. Nomographic analysis indicates that the position of phase equality for this strait is virtual and would lie in Cape Cod Bay beyond the east end of the canal, where A would equal 1.58 feet, that a node is present at the Railroad Bridge and that $R = 2$ and $\mu = 3$. When these values are

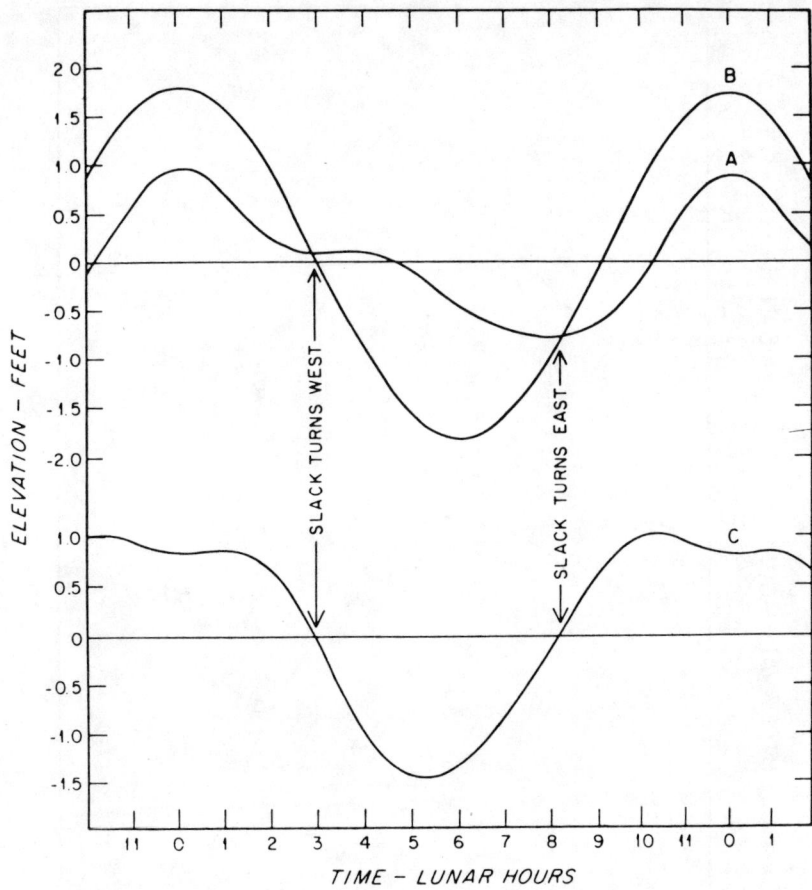

Fig. XII-6. Woods Hole Passage. Diagram of approximate tidal relations. A: Elevation of sea surface in Great Harbor due to M_2 and harmonics in Vineyard Sound. B: Elevation at Uncatena Island assuming harmonics to be negligible. C: Head between Uncatena Island and Great Harbor. Ordinate: Elevation in feet. Abscissa: Time in lunar hours relative to high water at Uncatena Island. (From the *Journal of Marine Research*, Vol. 12. With permission.)

introduced into the theoretical equations in Chapter II, the mean ranges and the high water intervals calculated agree with those predicted for positions on the canal (Figure XII-7). Slack water occurs in the canal one hour or more earlier than estimated by the theoretical equation which confirms the view that hydraulic currents are present. However, slack water does not occur simultaneously along the Canal, but is earlier at its east end, so it must be concluded that currents which accompany the waves which enter the Canal from opposite ends have a significant effect on the currents observed in the Canal. A study made prior to the present widening and deepening of the

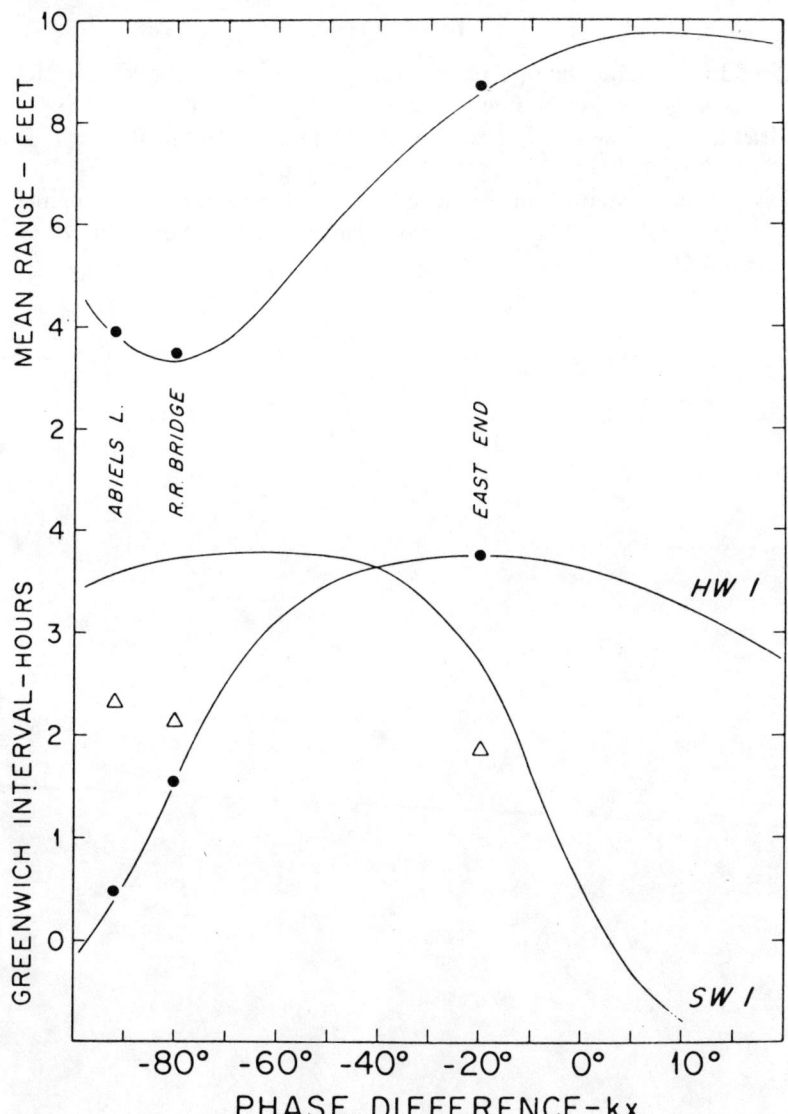

Fig. XII-7. Cape Cod Canal. Comparison as a function of kx of the theoretical values of mean range and the Greenwich intervals of high and slack water when $R = 2$ and $\mu = 3$ with their predicted values. (From the *Journal of Marine Research*, Vol. 36. With permission.)

Canal indicated that at that time also the flow in the Canal approximated more nearly to hydraulic than tidal conditions. The average minimum currents at the Railroad Bridge are 4 knots in the eastward direction and 4.5 knots in the westward direction.

References

Redfield, A. C. Interference phenomena in the tides of the Woods Hole region. Journal of Marine Research, 12, 121-140, 1953.

Redfield, A. C. The tide in coastal waters. Journal of Marine Research, 36, 255-294, 1978.

Panish, C. K. Hydraulics of the Cape Cod Canal. Report of U. S. Engineer Officer, Division Engineer, North Atlantic Division, New York, 29 pp. April 1933.

Chapter XIII

NEW YORK WATERS

The ocean tide enters the waters of New York between Sandy Hook and Rockaway Beach where the mean range has been increased in crossing the continental shelf to 4.6 feet and its crest arrives 0.32 hour after the moon's transit at Greenwich. Its principal passage into the region is across New York Lower Bay, through the Narrows to New York Upper Bay, past the Battery on Manhattan Island and thence up the Hudson River to Troy where a dam marks the limit of tide water. Along this course are a number of tributaries each with a characteristic tidal behavior.

Figure XIII-1 shows the relation of the time of high water to that of slack water as predicted for positions along the principal passage of the tide. Recalling that in a progressive wave slack water occurs one-quarter of the period (in the case of a semidiurnal wave 3.1 hours) after high water, and that in a reflected co-oscillation high water and slack water coincide at the position of reflection, the upper diagonal line in the diagram shows the relation to be expected in a progressive wave and the lower one that to be expected at the position of reflection. The predicted values for positions along the course vary between these extremes.

In New York Lower Bay, represented by Coney Island, the relation between the time of high water and that of slack water is nearly that to be expected in a reflected co-oscillation. The tide appears to result chiefly from the wave entering from the ocean and its reflection from the shores. In traversing the Narrows and crossing the Upper Bay, the relation changes greatly and on reaching the entrance to the Hudson River at the Battery, it becomes that of a progressive wave unmodified by reflection. This appears to be an example of the principle stated on page 6 that energy imparted to the water locally spreads to other regions as a progressive wave. In the lower reach of the Hudson River, between the Battery and the upper part of the Tappan Zee, at Ossining, the relation of high water to slack water is approximately that of a progressive wave. Above Ossining there is a long region of transition in which the interval of slack water gradually approaches that of

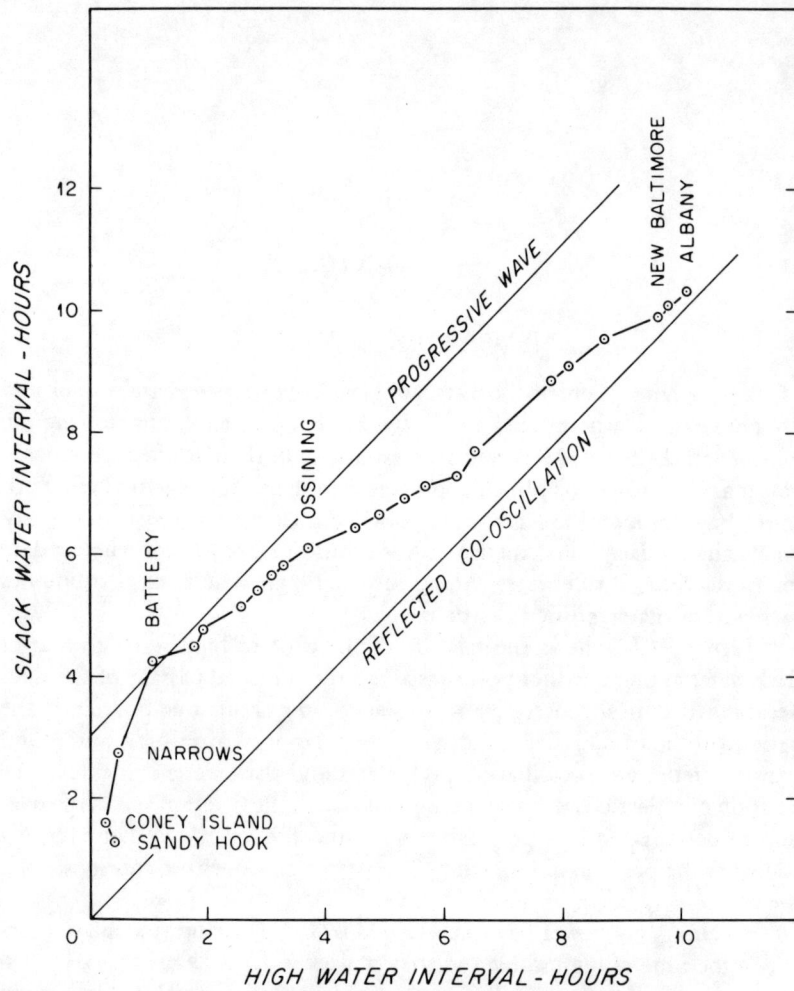

Fig. XIII-1. The relation of the time of high water and that of slack water between Sandy Hook and Albany. The diagonal lines indicate the relations in a progressive wave (above) and at the position of reflection (below).

high water until New Baltimore is reached. Above New Baltimore slack water and high water nearly coincide as they should near the position of reflection.

Figure XIII-2 is the nomogram for $R = 0$, a single progressive wave, on which are entered the predicted values of log η_H / η_{HO} and σt_H for positions on the lower Hudson River, after correcting for the asymmetry attributable to harmonics and, in this case, to the river flow and taking the tide at the Battery, where $A = 2.25$ feet, to be the origin of time and phase.

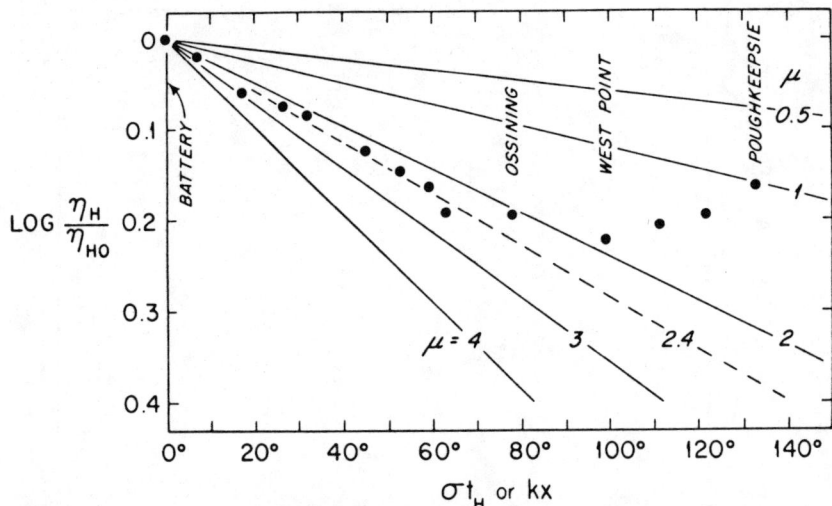

Fig. XIII-2. Nomogram for a progressive wave. Points indicate relations predicted at positions between the Battery and Poughkeepsie. They indicate that the tide is due to a progressive wave, not significantly influenced by reflection nearly up to Ossining below which $\mu = 2.4$. (From the *Journal of Marine Research*, Vol. 36. With permission.)

The predicted values fall about a line for $R = 0$ and $\mu = 2.4$ until Ossining is reached, above which they depart progressively. When these values are introduced into the theoretical equations, Chapter II, the values for the mean range and the interval of high water agree with those predicted while those for slack water are only slightly greater than those predicted (Figure XIII-3). The demonstration that a single progressive wave accounts for the tide in the lower reach of the Hudson River is of interest because this form of motion is assumed for the component waves in developing the theory of the tide in straits and embayments.

In the upper reach of the Hudson River between New Baltimore and Troy, the tide behaves as a reflected co-oscillation, reflected from the dam at Troy. Nomographic analysis indicates that $A = 1.175$, $R = 1$ and $\mu = 4$. When these values are inserted into the theoretical equations, Chapter II, the mean range and the time of high water agree with the predicted values between Troy and New Baltimore. The predicted time of slack water occurs more than one hour before that given in theory (Figure XIII-4). The discrepancy in the time of slack water may be attributed to the nontidal flow of the river. Slack water occurs when the current due to the incoming tide is equal to that of the river flow, not when the incoming tidal wave is equal to its reflectant. It is consequently earlier than that calculated by theory.

Between the upper part of the Tappan Zee and New Baltimore there is a transition region in which the behavior of the tide gradually changes from

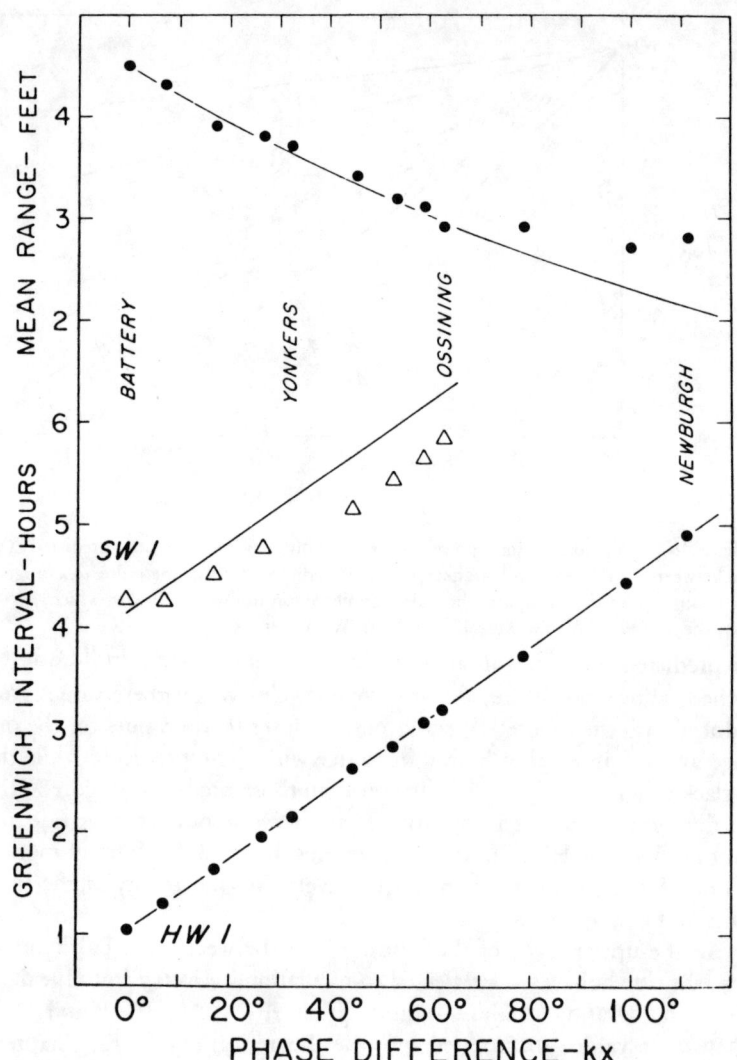

Fig. XIII-3. Lower Reach of Hudson River. Comparison as a function of kx of the theoretical values of mean range and the Greenwich intervals of high water and slack water when $A = 4.5$ ft., $R = 0$ and $\mu = 2.4$ with the predicted values. (From the *Journal of Marine Research*, Vol. 36. With permission.)

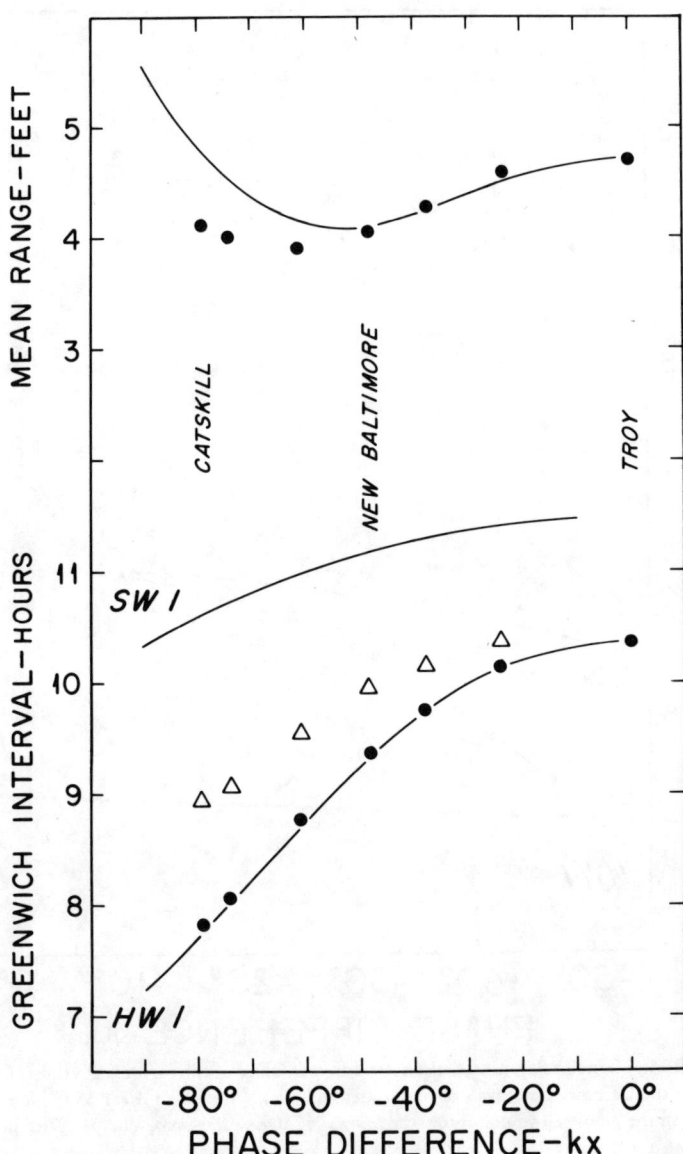

Fig. XIII-4. Upper Reach of Hudson River. Comparison as a function of kx of the theoretical values of mean range and the Greenwich intervals of high water and slack water when $A = 1.175$ ft., $R = 1$ and $\mu = 4$ with the predicted values. (From the *Journal of Marine Research*, Vol. 36. With permission.)

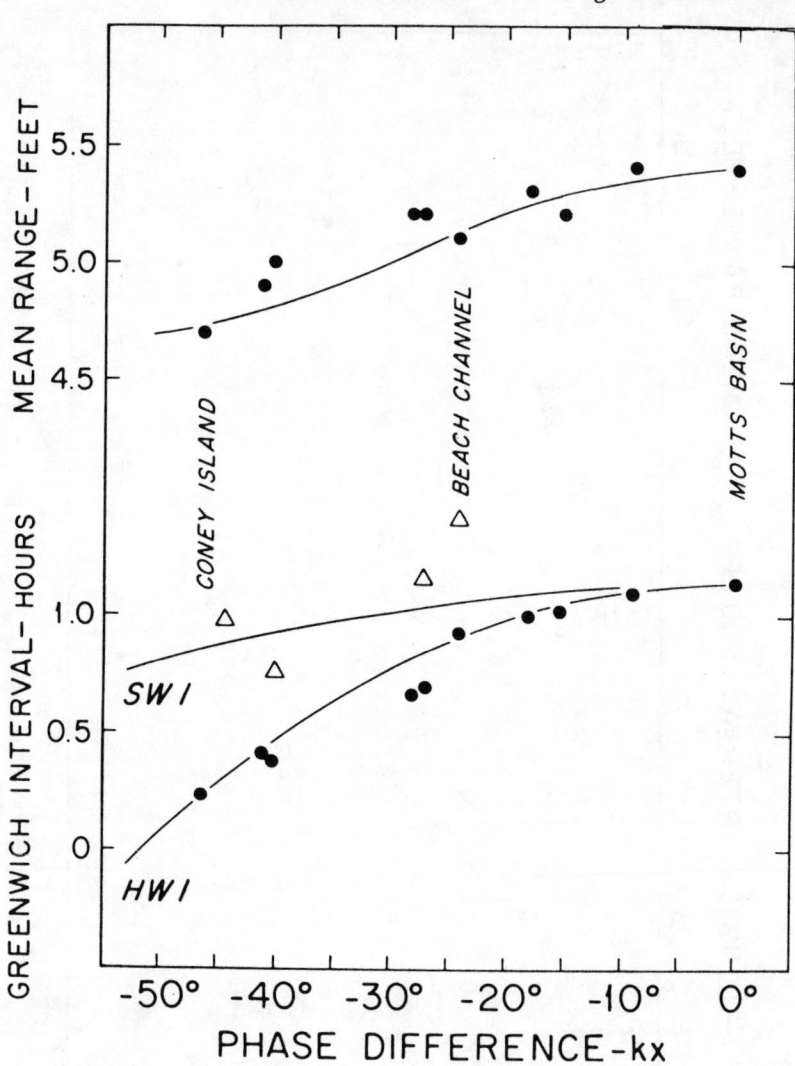

Fig. XIII-5. Jamaica Bay. Comparison as a function of kx of the theoretical values of mean range and the Greenwich intervals of high water and slack water when $A = 1.35$ ft., $R = 1$ and $\mu = 4$ with the predicted values. (From the *Journal of Marine Research*, Vol. 36. With permission.)

that of a progressive wave to that of a reflected co-oscillation. No satisfactory theoretical explanation can be given except that reflection evidently becomes less important as the river is descended.

The National Oceanic and Atmosphere Administration has issued Tidal Current Charts for New York Harbor which show the direction and

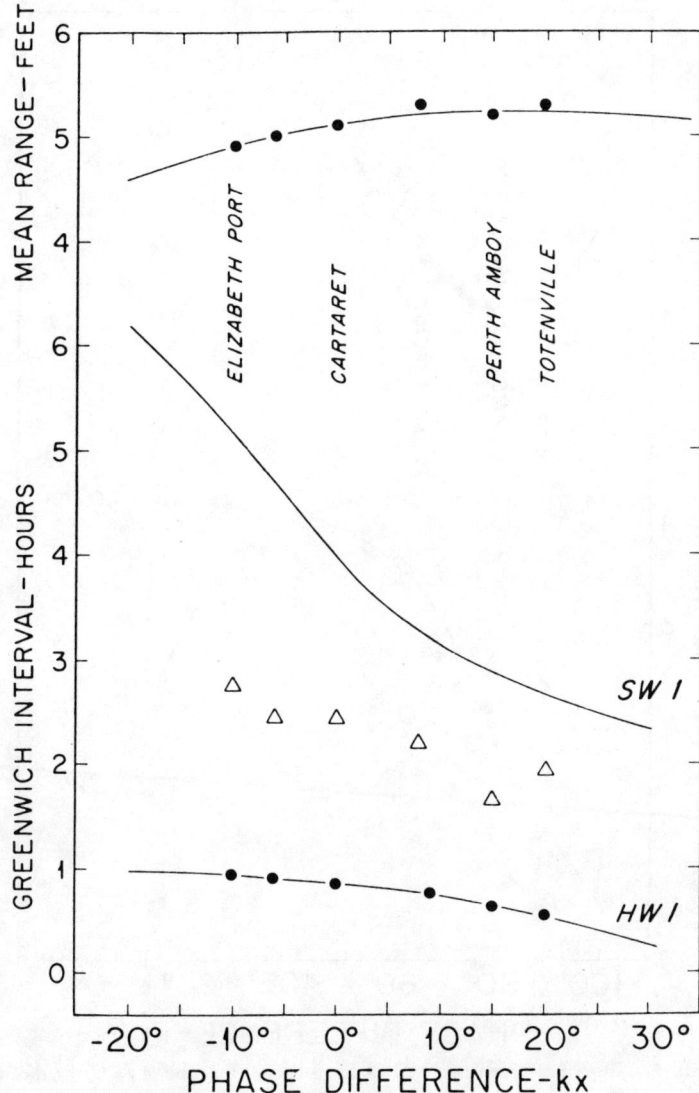

Fig. XIII-6. Arthur Kill. Comparison as a function of kx of the theoretical values of mean range and the Greenwich intervals of high water and slack water when $A = 0.85$ ft., $R = 2$ and $\mu = 3$ with their predicted values. (From the *Journal of Marine Research*, Vol. 36. With permission.)

strength of the currents at hourly intervals relative to high or low water at New York (the Battery).

Jamaica Bay is a tributary of New York Lower Bay which is entered through Rockaway Inlet. Nomographic analysis, taking Motts Basin as the

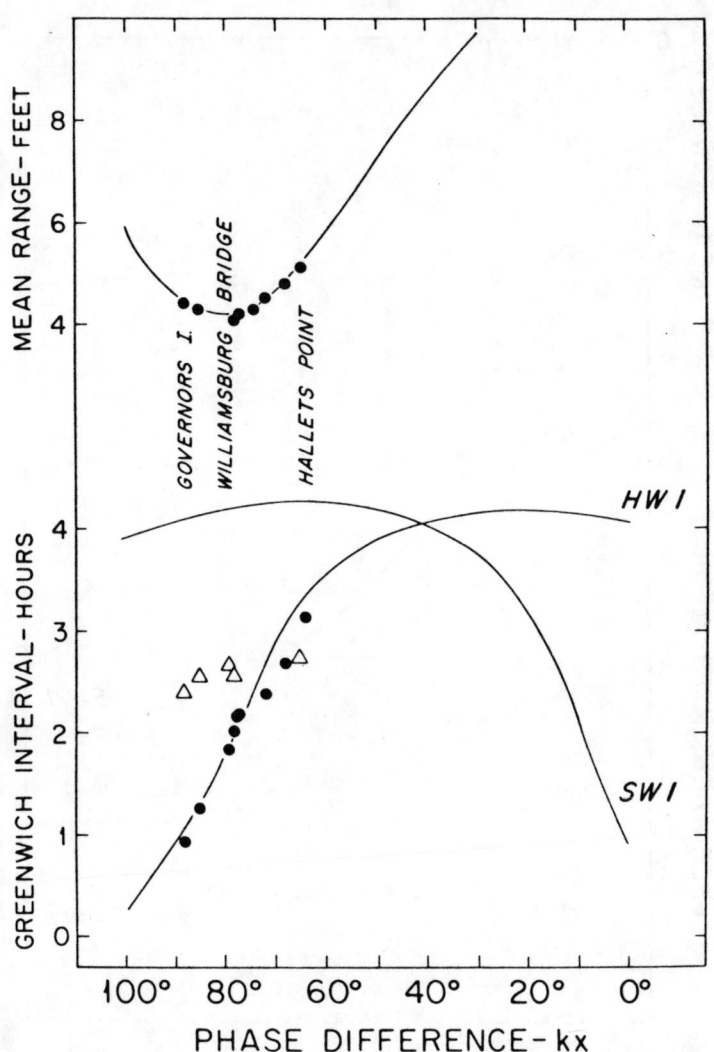

Fig. XIII-7. Lower Reach of East River. Comparison as a function of kx of the theoretical values of mean range and the Greenwich intervals of high water and slack water when $A = 1.97$ ft., $R = 2$ and $\mu = 3$ with their predicted values. (From the *Journal of Marine Research*, Vol. 36. With permission.)

position of reflection where $A = 1.35$, indicates that the tide is a reflected co-oscillation in which $R = 1$ and $\mu = 4$. When these values are introduced into the theoretical equations, Chapter II, the mean range and times of high water agree roughly with those predicted. There is no indication that hydraulic currents occur in Jamaica Bay or Rockaway Inlet (Figure XIII-5).

Raritan Bay is also a tributary of New York Lower Bay, lying west of Sandy Hook, but otherwise its outer limit is not defined by the topography. The mean range is slightly higher at the head of Raritan Bay than in the Lower Bay, being 5 feet at South Amboy. It is probably augmented by a reflected co-oscillation between the tide in the Lower Bay and its reflection from the shore. The changes in range and time of high water are so small that a meaningful nomographic analysis cannot be made.

The *Arthur Kill* is a strait separating Staten Island from the New Jersey mainland and connecting the head of Raritan Bay with the junction of Newark Bay and Kill van Kull and thence with New York Upper Bay. The channel is dredged through much of its width to a controlling depth of 35 feet. Nomographic anaylsis indicates that the position of phase equality is near Cartaret where $A = 0.85$ foot, $R = 2$ and $\mu = 3$. When these values are introduced into the theoretical equations, Chapter II, the mean range and the high water interval calculated agree closely with those predicted for positions along the strait (Figure XIII-6). Slack water, however, is predicted to occur much earlier than is indicated by theory, which may be interpreted to mean that hydraulic currents are present in the Arthur Kill. Slack water does not occur simultaneously in all parts of this strait, but is predicted to occur nearly an hour earlier in its southern than at its northern end. Hydraulic currents which arise from the strait must be considered to contribute significantly to the flow through this channel.

The *East River* is a strait separating Long Island from Manhattan Island and the mainland. It connects New York Upper Bay with the head of Long Island Sound. The relations of the range of tide and the time of high water show that the East River cannot be treated as a strait in which a single value of μ may be applied to all parts. However, by dividing the strait into three reaches each of which is somewhat similar in topography, but differs from the others, satisfactory nomograhic analyses may be made of each.

The *Lower Reach of the East River* extends from the Upper Bay at Governors Island to Hallets Point. The topography is simple, being complicated only by the presence of Welfare Island and the entrance of the Harlem River above it. Nomographic analysis indicates that the position of phase equality is virtual, being located above this reach where $A = 1.97$ feet, $R = 2$ and $\mu = 3$. When these values are introduced into the theoretical equations, Chapter II, the mean range and high water interval calculated agree satisfactorily with the predictions for positions along this reach (Figure XIII-7). Slack water occurs about 1.5 hours before that calculated and is nearly simultaneous at all positions. From this it may be concluded that hydraulic currents dominate the flow in this reach. At Welfare Island maximum currents having an average velocity of 4.7 knots are predicted during the southward ebb and 3.8 knots during the flood. A node occurs below the

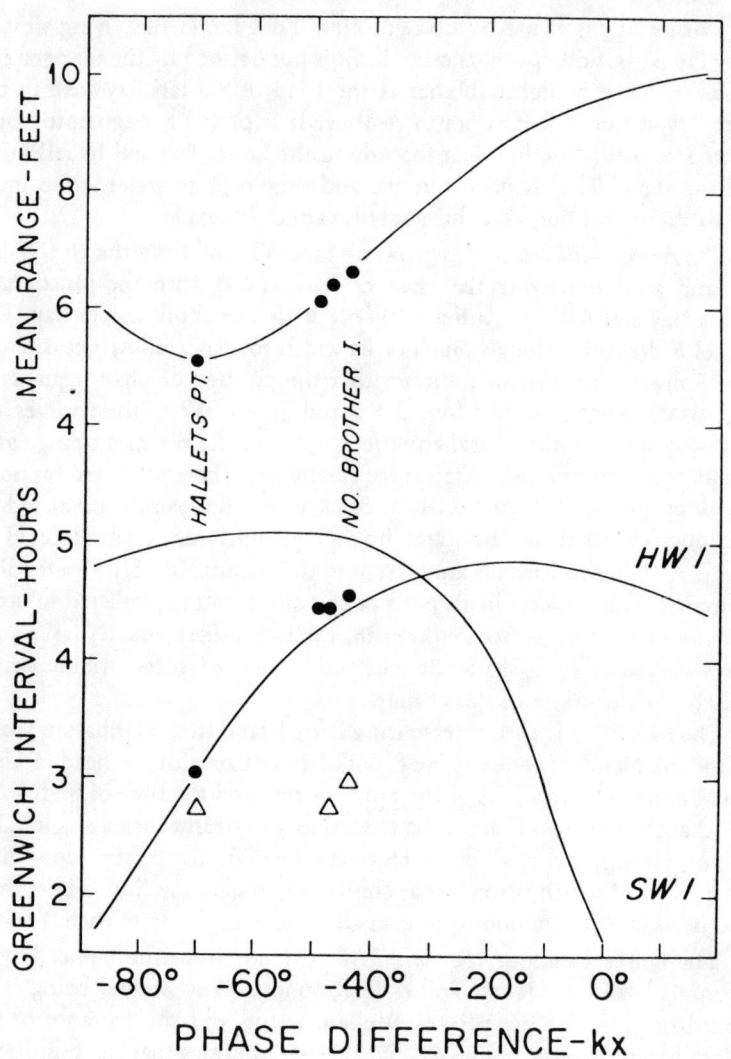

Fig. XIII-8. Middle Reach of East River. Comparison as a function of kx of the theoretical values of mean range and Greenwich intervals of high and slack water when $A = 1.61$ ft., $R = 2$ and $\mu = 4$ with their predicted values. (From the *Journal of Marine Research*, Vol. 36. With permission.)

Williamsburg Bridge where the mean range is reduced to 4.1 feet.

The *Middle Reach of the East River* extends from Hallets Point to North Brother Island. This reach is relatively shallow and its topography is irregular, being broken by the presence of several islands. Nomographic

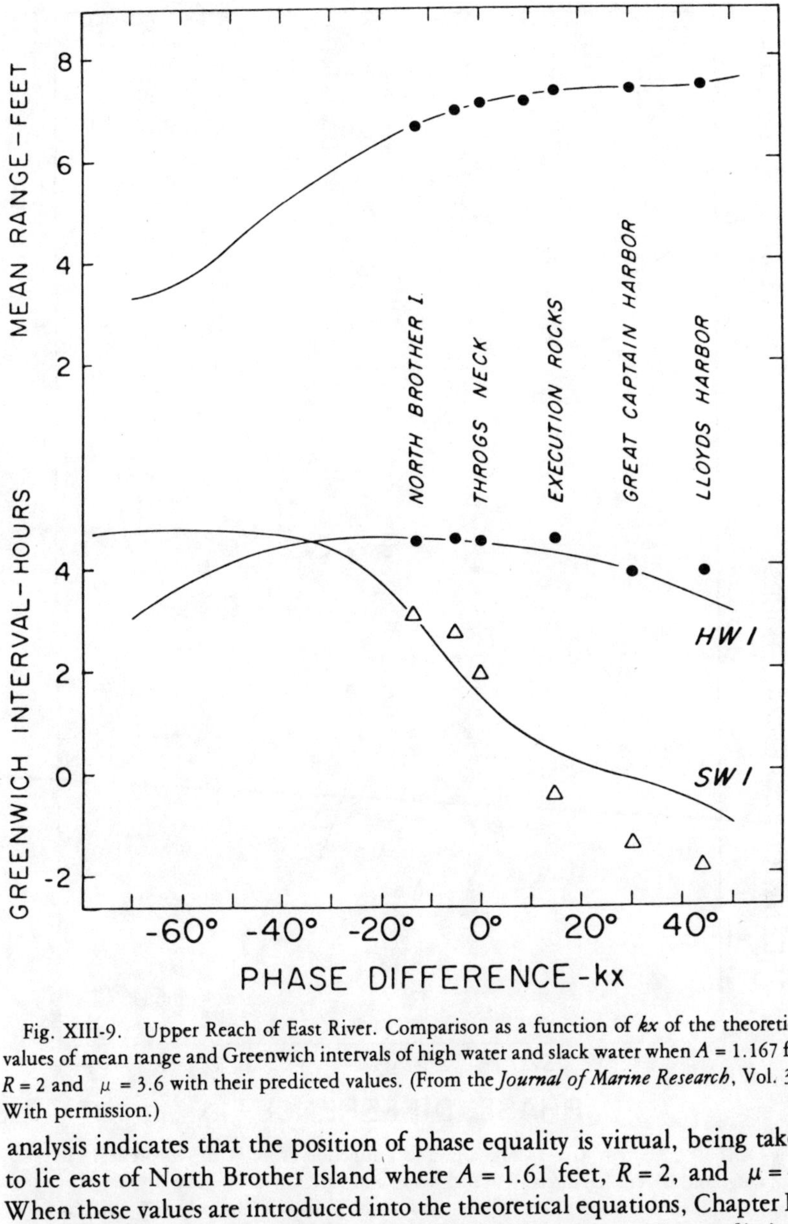

Fig. XIII-9. Upper Reach of East River. Comparison as a function of *kx* of the theoretical values of mean range and Greenwich intervals of high water and slack water when $A = 1.167$ ft., $R = 2$ and $\mu = 3.6$ with their predicted values. (From the *Journal of Marine Research*, Vol. 36. With permission.)

analysis indicates that the position of phase equality is virtual, being taken to lie east of North Brother Island where $A = 1.61$ feet, $R = 2$, and $\mu = 4$. When these values are introduced into the theoretical equations, Chapter II, the mean range and the high water interval agree well with the predictions for positions along this reach (Figure XIII-8). A node occurs at Hallets Point where the mean range is reduced to 5.1 feet. Slack water is predicted to occur about 2 hours before it is indicated by the theoretical equations which

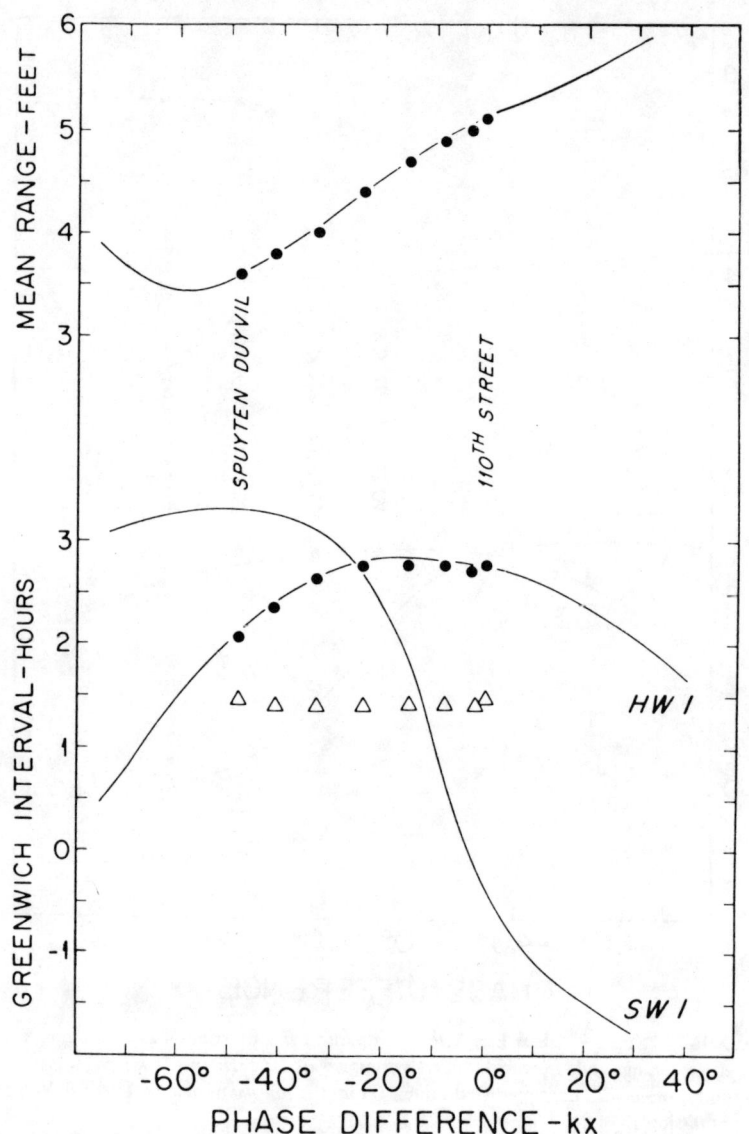

Fig. XIII-10. Harlem River. Comparison as a function of kx of the theoretical values of mean range and Greenwich intervals of high water and slack water when $A = 0.85$ ft., $R = 2$ and $\mu = 5$ with their predicted values. (From the *Journal of Marine Research*, Vol. 36. With permission.)

indicates that the current in the Middle Reach is predominantly hydraulic. The maximum velocity of the average current at the Brother Islands is 2.5 knots during the eastward flood and 1.8 knots during the westward and increases westward to the offing of Halletts Point where maximum currents of 3.4 and 4.6 knots are predicted.

The *Upper Reach of the East River* extends from North Brother Island to Throgs Neck, beyond which the channel widens into the head of Long Island Sound with little change in the range or the time of high water. The topography of this reach is irregular, several shallow bays being tributary to the main channel. Nomographic analysis, taking Throgs Neck to be the position of phase equality where $A = 1.167$ feet, indicates that $R = 2$, $\mu = 3.6$. When these values are introduced into the theoretical equations, Chapter II, the mean range and the intervals of high water and slack water all agree with the predictions for positions along this reach (Figure XIII-9). There is no evidence that significant hydraulic currents occur in the Upper Reach of the East River. The maximum average current predicted for this reach is 1.7 knots at Hunts Point and is probably not greater than that due to the interference of waves which enter the Reach from either end.

Considering the East River as a whole the current is predominantly hydraulic in nature between New York Upper Bay and North Brother Island. Above North Brother Island it appears to be due entirely to the interference of waves entering from the Middle Reach and from the head of Long Island Sound. The attenuation undergone by the tidal waves traversing the East River differs in its several parts as the result of differences in the topography. It is greatest in the Middle Reach ($\mu = 4$) and least in the Lower Reach ($\mu = 3$). Hell Gate is the name locally applied to the region about Halletts Point. It is so called because of the great strength of the currents there and the difficulties they impose on navigation.

The *Harlem River* is a strait separating Manhattan Island from the mainland and connecting the Hudson River at Spuyten Duyvil with the East River above Welfare Island. Nomographic analysis, taking the position of phase equality to be at 110th Street where $A = 0.85$ foot, indicates that $R = 2$ and $\mu = 5$. When these values are introduced into the theoretical equations, Chapter II, the mean ranges and the intervals of high water calculated agree with those predicted for positions along the Harlem River (Fig. XIII-10). Slack water is predicted to occur practically simultaneously in all parts of the strait from which it may be concluded that the current is predominantly hydraulic in nature (Figure V-1).

References

Marmer, H. A. Tides and Currents in New York Harbor. U. S. Deptartment of Commerce, Coast and Geodetic Survey, Special Publication No. 111, 198 pp., 1935.

Redfield, A. C. The tide in coastal waters. Journal of Marine Research, 36, 255-294, 1978.

Appendix A

OBSERVATIONS AND EXPERIMENTS ON WAVES

The following observations and experiments will make the account of progressive waves given in Chapter II more real to the reader who will take the trouble to try them.

Observations

The next time it rains, observe a shallow pool such as collects in a depression of a roadway. Each raindrop produces a circular wave which moves with increasing diameter away from the point where the drop falls. The energy imparted to the water by the falling drop spreads from its position as a progressive wave. As the circumference of the circular wave increases, the energy of each part decreases and, as a result, the wave soon becomes invisible.

When two drops fall near one another, the circular waves produced by each meet and pass on as though the other did not exist. The waves behave as though independent of one another. At any point the elevation of the water relative to the still-water level is the sum of the elevation of each wave.

Experiments

Secure an aquarium about 20 inches long, 10 inches wide and at least 10 inches deep. Place water to a depth of ½-inch in the tank. Lift an end as in A (Figure A-1), then let it return to the horizontal as in B. The water will flow as a *progressive wave* across the tank as in C. The front of the wave is steep and turbulent because of the shallowness, as is that of the *tidal bore* described in Chapter VI. On reaching the end of the tank, the water piles up against the barrier and the wave then moves back in the direction from which it has come as in D. The wave has been *reflected* from the barrier. Allow the wave to flow back and forth until it has subsided. The wave has been *attenuated*.

The development of a *standing wave* by the reflection of a progressive wave is demonstrated by increasing the depth of the water. With a depth of 2 or 3 inches the wave assumes a symmetrical sinusoidal form and is no

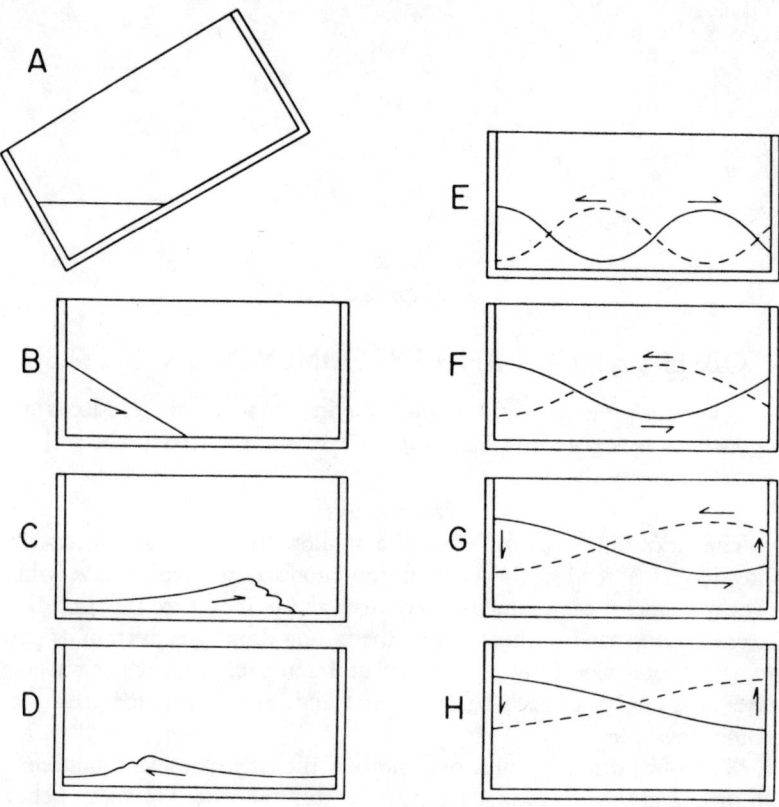

Fig. A-1. Development of a standing wave from a progressive wave and its reflection.
A to C. Formation of a progressive wave.
D. Its reflection
E to G. The timing of the progressive wave and its reflection is modified as the depth increases to form at
H. A standing wave.

longer turbulent as it was in very shallow water (see E). Its progression back and forth remains evident. With a depth of 5 or 6 inches, the water surface tilts up and down rhythmically on either side of the middle of the tank. A wave may be seen to move back and forth across the surface as in F. With a depth of 8 inches, the progression of the disturbance disappears. The water has a flat surface which moves up and down on opposite sides of a *nodal line* at the middle of the tank as in H. The motion has become that of a *standing wave*.

If particles suspended in the water are observed, they may be seen to move back and forth in the direction toward which the water is rising. Their horizontal excursion is greatest at the nodal line decreasing to zero toward

Observations and Experiments on Waves 105

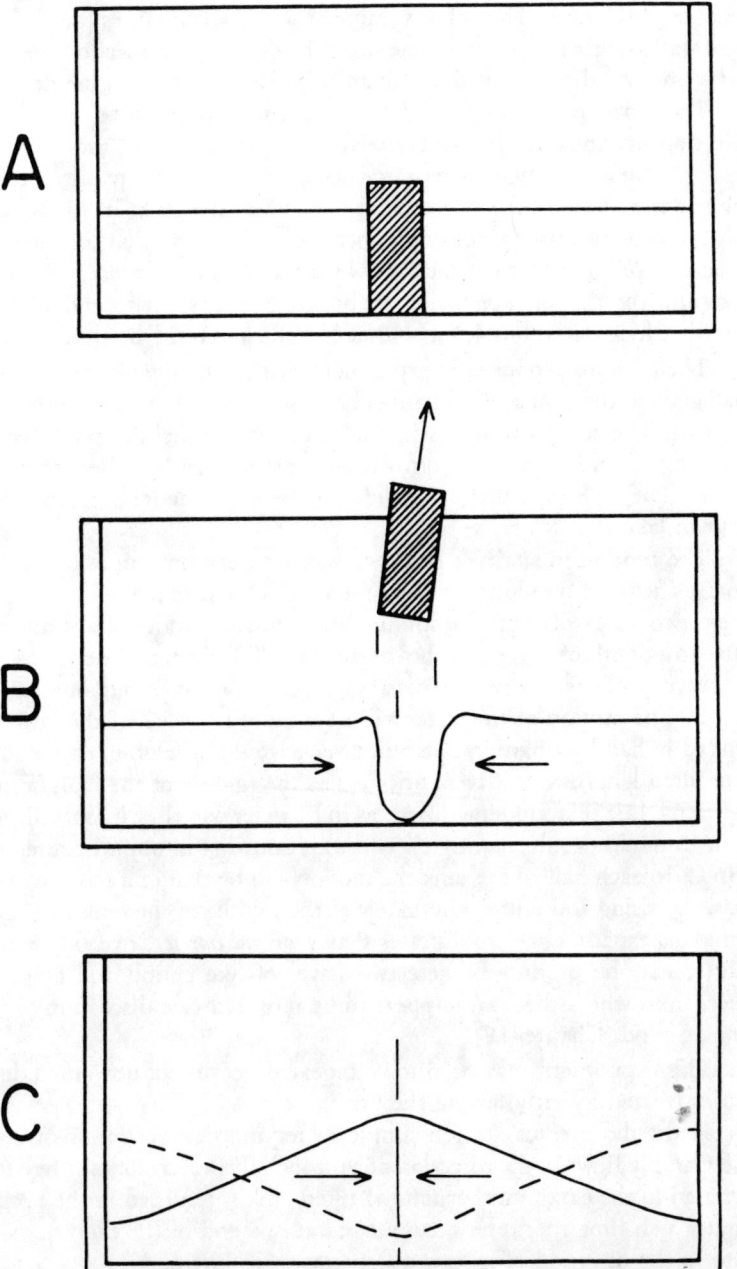

Fig. A-2. An antinode which acts as an apparent barrier is produced where two waves of the same period and amplitude meet.

the ends of the tank. The entire volume of water which causes the surface to rise or fall at either end of the tank must flow through the section under the nodal line. As the ends of the tank are approached, this volume decreases.

The development of a standing wave from a progressive wave and its reflection depends on the correct relation of the length of the tank to its depth. If the experiment is repeated measuring the time required for the water to rise and fall, say 10 times, at one end of the tank, it will be found that the time decreases as the depth increases. This shows that the rate of advance of a progressive wave increases with the depth of water. In a 20-inch aquarium the rate of progression is about 1.5 inches per second when the water is 1 inch deep, but 4.3 inches per second when the depth is 8 inches.

The motions produced in experiments with a small tank are not strictly analogous to those in a tidal channel because the latter is open to the sea at one end. The successive waves which enter the channel are produced and timed by the tide entering from without rather than by reflection from a barrier. The tank experiments provide a more exact model of *seiches* which occur in lakes.

The motion in straits, where two waves enter from opposite ends, are more difficult to reproduce in a simple tank. One case may be represented where two waves of similar amplitude meet in the middle of a strait. Place water to a depth of 2.5 to 3 inches in the tank. Tilt the tank sidewise, so that the wave produced moves across the width which is about one-half the length. The motion of the water will be a standing wave. If the tank were divided in half by a barrier, a standing wave would develop in each half if it were tilted lengthwise. Place a brick across the middle of the tank (as in A, Figure A-2). If it is suddenly lifted, as in B, water will flow into its displacement from either side, setting up two waves moving in opposite directions, as in C. In each half of the tank the motion will be that of a standing wave, the water rising and falling alternately in the middle of the tank and at each end. This experiment demonstrates that a *virtual barrier*, hydrodynamic in nature, may be produced where two waves of like period and amplitude meet, from which the waves appear to be reflected (see discussion of Long Island Sound, Chapter IV).

This experiment may need to be repeated because of unwanted disturbances created in withdrawing the brick.

Hydraulic streams are too simple to require demonstration since the water simply flows in the direction down which the water slopes. They are illustrated in the experiment when the tilted tank is returned to the horizontal at which time the water accumulated at one end of the tank flows outward in the direction of its down slope, as in B (Figure A-1).

The effect of the earth's rotation on the motion of water is too little to be demonstrated in a small tank. It may be reproduced by substituting cen-

Observations and Experiments on Waves

trifugal force for the effect of the earth's rotation. Place water in a large cylindrical jar or carboy to a depth equal to about half its diameter, tilt the vessel, give it a twirling motion, then return it to the upright position. The motion developed is peculiar. The surface is practically flat but is tilted (Figure A-3). When viewed at a point on the circumference, the elevation of the water moves up and down as though a sinusoidal wave were progressing around the wall of the vessel. When viewed along a diameter, the surface appears to oscillate up and down about a node where there is no vertical movement, as though a standing wave were present. Figure A-3 shows the momentary slope of the surface as it rotates and the resulting pattern of cotidal and corange lines. the central diagram in Figure A-3 shows the resulting pattern of radial cotidal lines and of circular corange lines in which the range increases outward from an amphidromic point where there is no vertical motion.

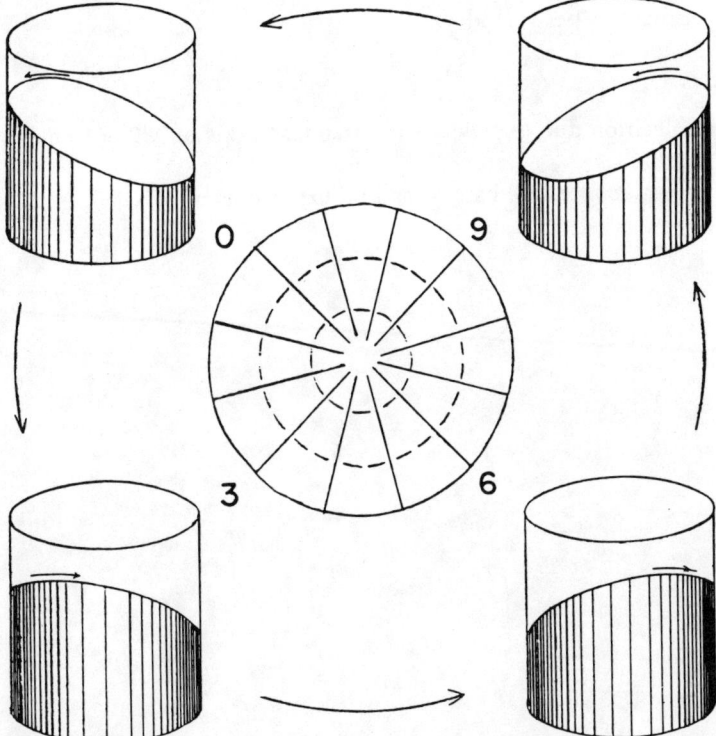

Fig. A-3. Diagram of an amphidromic system of motion produced in a vessel by rotation. The outer figures show the slope of the water surface at successive times equal to ¼ period of rotation. The central diagram shows the successive radial cotidal lines at intervals of 1/12 of the period of rotation and the circular corange lines which increase from an amphidromic point at the center toward the circumference of the vessel.

Appendix B

Conversion Factors for Units of Measurement and Useful Constants

Conversion factors

1 foot = 30.48 centimeters or 0.3048 meters

1 fathom = 6 feet or 1.8288 meters

1 nautical mile = 6080.2 feet or 1.152 statute miles or 1.853 kilometers or one minute of arc on the earth's surface at the equator

1 knot = 1 nautical mile per hour or 1.152 statue miles per hour or 51.48 centimeters per second

Constants

The acceleration due to gravity has a standard value of 980.665 cm/sec^2 or 32.174 ft/sec^2.

The value of e, the base of natural logarithms, is 2.72828.